Basic Naval Architecture

Philip A. Wilson

Basic Naval Architecture

Ship Stability

 Springer

Philip A. Wilson
Faculty of Engineering and the Environment
University of Southampton
Southampton
UK

ISBN 978-3-319-89212-2   ISBN 978-3-319-72805-6   (eBook)
https://doi.org/10.1007/978-3-319-72805-6

This Springer imprint is published by the registered company Springer International Publishing AG part of Springer Nature
The registered company address is: Gewerbestrasse 11, 6330 Cham, Switzerland

Acknowledgements

This book is dedicated to the staff and students of Ship Science who over the many years have been a fruitful source of inspiration to write this book. The work is based upon lecture notes for what was formerly the courses called Ship Studies and in later years is Basic Naval Architecture.

In particular I would thank my former colleagues David Cooper, John Wellicome and Penny Temarel.

Of course I could not have found time to write this work without the help and support of my wife Hilary and our four children, now fully grown up and who have fled the nest, Richard, Thomas, Jennifer and Hugh. The cat Mir was no help whatsoever!

Southampton, UK   Philip A. Wilson
May 2017

Contents

1 Introduction to Naval Architecture 1
   1   Introduction to Maritime Transportation 1
   2   World Seaborne Trade........................... 2
   3   Survey of Maritime Vessels....................... 3
   4   A Global View of Ship Design—the Design
       Requirements................................ 5
   5   The Design Cycle 5
   6   The Geometry of a Ship's Hull..................... 6
   7   Comparative Design Parameters 8
       7.1 Deadweight Coefficient..................... 9
       7.2 Slenderness Coefficients 9
       7.3 Fineness Coefficients...................... 10
       7.4 Speed Parameters 11
       7.5 Design Trend Lines....................... 11
       7.6 Displacement Mass and Weight 13
   8   Terms Used to Define the Midship Section.............. 13
   9   Summary.................................. 14

2 Basic Properties............................... 15
   1   Mass, Weight and Moments of Weight................. 15
   2   Moment of Weight 15
   3   Transfer of Weight—Equivalent Forces and Weights 16
   4   Centres of Gravity............................ 17
   5   Summation Notation 17
   6   Estimation of Point of Balance 18
   7   The Effect of Rotation on Moment Acting 19
   8   General Expressions for Centre of Gravity 19
   9   Example Calculations of Centre of Gravity.............. 21
   10 Summary.................................. 22

3 Equilibrium and Stability Concepts for Floating Bodies 23
   1   Pressures in a Uniform Incompressible Fluid at Rest   , , , , . 23
   2   Pressures on a Closed Surface *S* at Rest in a Fluid also
       at Rest.................................... 23

3   Archimedes Principle.   24
4   Calculating Force of Buoyancy   25
5   Equilibrium and Stability of Floating Bodies.   25
       5.1   General Definitions   25
       5.2   Stability of a Submerged Body   27
       5.3   Stability of a Floating Body   27
6   Summary   28

4   Calculating Volumes and Centres of Buoyancy   29
1   Integration as a Limit of Summation   29
2   Areas and Centres of Area of Laminæ   32
3   A Simple Example   34
       3.1   Rectangular Lamina $(n = 0)$ 35
       3.2   Triangular Lamina $(n = 1)$ 35
       3.3   Parabolic Lamina I, $(n = 1/2)$ 36
       3.4   Parabolic Lamina II, $(n = 2)$.   37
4   Summary   37

**5   Further Comments on Displacement Volume and Centre
       of Buoyancy**   39
1   Calculation of Displacement and Centre of Buoyancy.   39
2   Calculation of Sectional Area   41
3   Calculation of Waterplane Area and Centroid   41
4   Introduction to Changes of Draught (Parallel Sinkage)
       and Trim   42
5   Movement of LCB Due to a Small Change of Trim   43
6   Longitudinal Second Moments of Area and Parallel Axis
       Theorem   45
7   Formulæ for LCB Shift and Longitudinal Metacentre   46
8   Movement of Centre of Buoyancy Due to Small Heel Angle   47
9   Second Moments of Area of Simple Laminæ   49
       9.1   Rectangular Lamina   49
       9.2   Typical Ship Waterplane.   51
       9.3   Mathematically Defined Waterplane   51
       9.4   Circular Lamina   53
10   Summary   54

6   Numerical Integration Formulæ.   57
1   Trapezoidal Rule   58
2   Simpson's First Rule   59
       2.1   Example   61
3   Simpson's Second Rule   63
4   5, $+8$, -1 Rule   64

7 Problems Involving Changes of Draught and Trim 67
  1   The Position so far 67
  2   Moment to Change Trim 68
  3   Trimmed Draughts 69
  4   Adding Mass to a Vessel 70
        4.1 Example 1 70
        4.2 Example 2 71
  5   Moving from Freshwater to Salt Water 72
        5.1 Example 3 73
  6   Docking a Vessel Trimmed by the Stern 74
        6.1 Example 4 74
  7   Variation of Hydrostatic Particulars with Draught 76
  8   The Inclining Experiment 77
        8.1 Purpose 77
        8.2 Method 77
        8.3 Precautions to Observe 78
        8.4 Measurements of Draught 79
        8.5 Corrections to Lightship 79

8 Transverse Initial Stability Topics 81
  1   Righting and Heeling Moments at Small Angles 81
  2   Metacentric Height Diagram for a Rectangular Box 83
  3   Stability of a Uniform Square Sectioned Log 85
        3.1 Log Floating with One Face Horizontal 85
        3.2 Log Floating with One Diagonal Horizontal 85
  4   Morrish's Formula for KB 88
  5   Munro-Smith Estimate of BM_T 89
  6   Initial Estimate of Ship Moulded Beam 91
  7   Losses of Transverse Stability -*Virtual* Centre Gravity
        Problems 92
        7.1 Suspended Weights 92
        7.2 Liquid Free Surfaces 93
        7.3 Stability Losses Due to Grounding or Docking 96
  8   Summary 98

9 Wall-Sided Formula and Applications 99
  1   Wall-Sided Formula 99
  2   Application to Transverse Movement of Weight 102
  3   Angles of Loll 103
  4   Summary 105

10 Large Angle Stability 107
  1   The Righting Lever GZ Curve 107
  2   Factors Affecting the GZ Curve 108
        2.1 Height of Centre of Gravity 108

	2.2	Increasing Beam	109
	2.3	Increasing Freeboard	110
	2.4	Watertight Superstructure	110
3	The Calculation of Righting Lever Curves		112
	3.1	Storage of Section Data	112
	3.2	Properties of a Full Section at an Angle of Heel	113
	3.3	Integrated Properties of Immersed Volume	114
	3.4	The Calculation of GZ	114
	3.5	Varying Draught and Trim	115
	3.6	Cross Curves Calculation Mode	115
4	Dynamical Stability		116
	4.1	Basic Concepts	117
	4.2	Application to Ships	118
	4.3	Response to Suddenly Applied Moments	119
	4.4	Stability Criteria	120
5	Summary		121

11 Flooding Calculations 123
   1   Definitions Used in Subdivision 123
       1.1   Bulkhead Deck 123
       1.2   Margin Line 124
       1.3   Compartment Permeability (μ) 124
       1.4   Floodable Length 124
   2   Added Weight and Lost Buoyancy Calculation Methods 125
       2.1   Added Weight Method 125
       2.2   Lost Buoyancy Method 125
   3   Flooding to a Specified Waterline 126
       3.1   Constructing a Floodable Length Curve 127
   4   Flooding a Specified Compartment 127
   5   Corrections for Sinkage and Trim 129
   6   Example: Added Weight Calculation 129
   7   Example: Lost Buoyancy Method 131
   8   Summary 133

12 End on Launching and Launching Calculations 135
   1   Ground Way Geometry 136
       1.1   Straight Ways 136
       1.2   Cambered Ways 136
   2   Launching Calculations 137
       2.1   Prior to Stern Lift 137
       2.2   Post Stern Lift 138
       2.3   Launch Curves 138
   3   Summary 139

**13 Stability Assessment Methods (Deterministic
     and Probabilistic)** 141
   1 Background 141
      1.1 IMO 141
      1.2 Ship Stability Developments......................... 141
      1.3 History of the Development of the Probabilistic
           Methodology... 144
   2 Damage Stability Calculations............................. 145
      2.1 Damage Extent 146
      2.2 Requirements 147
      2.3 Probabilistic Damage Stability 148
      2.4 Probabilistic Concept 150
      2.5 Excerpt from ANNEX 22 of SOLAS 152
   3 Detailed Regulations According to SOLAS 2009 154
      3.1 Subdivision Length................................. 154
      3.2 Calculation Method................................. 156
      3.3 Longitudinal Subdivision 157
      3.4 Regulation Definitions.............................. 158
      3.5 Light Service Draught.............................. 158
      3.6 Draught and Trim.................................. 158
      3.7 Required Subdivision Index R 159
      3.8 Attained Subdivision Index A 161
   4 Permeability... 168
   References.. 169

14 Second Generation Stability Methodology 171
   1 Introduction 171
   2 The IMO Second Generation Intact Stability Criteria 172
      2.1 Parametric Roll.................................... 172
      2.2 Pure Loss of Stability 178
      2.3 Dead Ship Stability................................. 180
      2.4 Excessive Acceleration 182
   3 Direct Stability Assessment (DSA) 184
      3.1 General Requirements............................... 184
      3.2 Parametric Roll and Excessive Acceleration 184
      3.3 Pure Loss of Stability 185
      3.4 Surf-Riding and Broaching-to........................ 185
      3.5 Dead Ship Condition............................... 185
   References.. 186

15 Examples 187
   1 Examples 1 187
   2 Examples 2 188
   3 Examples 3 189

4   Examples 4 191
5   Examples 5 192
6   Examples 6 194
7   Examples 7 195
8   Examples 8 197
9   Examples 9 198
10   Examples 10 199
11   Examples 11 200
12   Examples 12 201
13   Examples 13 202

List of Figures

Chapter 1
Fig. 1   Body plan (with kind permission of Dr. John Wellicome).   7
Fig. 2   Projection of hull lines (with kind permission
           of Dr. john Wellicome)   8
Fig. 3   Waterlines   11
Fig. 4   Section and waterline shapes   12
Fig. 5   Design trend lines, volume coefficient with Froude number   12
Fig. 6   Design trend line block coefficient variation
           with Froude number   13

Chapter 2
Fig. 1   Definition of moment   16
Fig. 2   Addition of equal forces at a singular point.   16
Fig. 3   Forces replaced by a couple plus force   16
Fig. 4   Moment of a couple   17
Fig. 5   Multiple forces added for equivalent moment on system.   17
Fig. 6   Generalisation to large number of point masses.   18
Fig. 7   2-D rotation of axis system   19
Fig. 8   3-D rotation of axis system   20

Chapter 3
Fig. 1   Pressure forces on the base of a vertical cylinder   24
Fig. 2   Pressure forces on a three dimensional body   24
Fig. 3   Stability conditions and types   26
Fig. 4   Stability of a fully submerged body.   27
Fig. 5   Effects of positions of centre of buoyancy and gravity   27

Chapter 4
Fig. 1   Sectional area curve.   30
Fig. 2   Integration as limit of summation   30
Fig. 3   Integration consider as a limit of summation of areas   31
Fig. 4   Integration with variable upper and lower limits   32

Fig. 5   A simple laminar with edges approximated by a generalised
          power of x...   34
Fig. 6   Area and centre gravity calculation for a triangular lamina......   35
Fig. 7   Area of a blunt-nosed parabolic laminar...................   36
Fig. 8   Area of pointed nosed parabolic laminar...................   37

Chapter 5
Fig. 1   Volume estimation by approximation of ship by longitudinal
          sections...   40
Fig. 2   Volume approximation of ship by using waterplane areas.......   40
Fig. 3   Section at distance x from amidships......................   41
Fig. 4   Waterplane area estimation from offset data.................   42
Fig. 5   Definition of parallel sinkage............................   42
Fig. 6   Trim change on waterline...............................   43
Fig. 7   Calculation of trimmed LCF...........................   44
Fig. 8   Definition of longitudinal metacentre M_L................   47
Fig. 9   Effects of heel assuming wall sided ship section.............   48
Fig. 10   Transverse metacentre definition.......................   49
Fig. 11   Calculation of J_T for a rectangular laminar.................   50
Fig. 12   Typical ship waterplane used for calculation of J_T............   51
Fig. 13   Mathematically defined waterplane......................   52
Fig. 14   Second moment of area, J_T and J_L for a circular laminar.......   53

Chapter 6
Fig. 1   Function evaluated at points spaced at even intervals..........   58
Fig. 2   Trapezoidal rule....................................   59
Fig. 3   Simpson's first rule.................................   60
Fig. 4   Simpson's second rule for even-spaced intervals.............   63
Fig. 5   Simpson's third rule.................................   65

Chapter 7
Fig. 1   Parallel sinkage....................................   68
Fig. 2   Trim about LCF....................................   68
Fig. 3   Moment to change trim...............................   69
Fig. 4   Trimmed draughts at perpendiculars......................   70
Fig. 5   Example of addition of mass on a vessel...................   71
Fig. 6   Ship being brought into dry dock........................   74
Fig. 7   Inclined section....................................   78
Fig. 8   Draught measurements...............................   79

Chapter 8
Fig. 1   Roll restoring moment................................   82
Fig. 2   Moment of inclining masses...........................   82
Fig. 3   Position of B,G and M_T.............................   82

Fig. 4   Calculation of M_T for a rectangular box 83
Fig. 5   Variation of box parameters with KM_T 84
Fig. 6   Stability range for a square sectioned log 85
Fig. 7   Square sectioned log floating on diagonal with shallow
           waterline 85
Fig. 8   Square sectioned log floating on diagonal with deep waterline ... 86
Fig. 9   Stability range of square sectioned log floating on a diagonal
           waterline 87
Fig. 10   Approximation of waterplane area using linear variations
           used in Morrish's method 88
Fig. 11   Weights suspended by a crane. 92
Fig. 12   A derrick under control booms or vangs 93
Fig. 13   Derrick under full control 93
Fig. 14   Free surface effects 94
Fig. 15   Reduction of free surface area in a vertical hopper 94
Fig. 16   Subdivision longitudinally in cargo tanks 95
Fig. 17   Change of G due to replacement of sea water for oil
           in fuel tanks. 95
Fig. 18   Free surface loss in partially filled tanks 96
Fig. 19   Stability loss due to grounding or docking 97
Fig. 20   Shores used in dry docking 98

Chapter 9
Fig. 1   Heeled ship section for wall-sided calculation 100
Fig. 2   Calculation of centre of buoyancy for heeled ship sections 101
Fig. 3   Explanation of calculation of the position of B 101
Fig. 4   Movement of mass across ship section 103
Fig. 5   Angle of loll 104

Chapter 10
Fig. 1   Righting moments on heeled vessel. 108
Fig. 2   Righting moment curve. 108
Fig. 3   Effects of increasing G on GZ. 109
Fig. 4   Effect of position of G on GZ curve 109
Fig. 5   Increasing beam and its effect on GZ 110
Fig. 6   Increasing freeboard 110
Fig. 7   Effect of increase of freeboard on GZ 111
Fig. 8   Heeled ship section 111
Fig. 9   Stability curve with/without integral superstructure 111
Fig. 10   Digitisation of ship section 112
Fig. 11   Calculation of heeled section data from upright Bonjean
           curves 113
Fig. 12   Detail of how to calculate the CG for heeled ships 115

Fig. 13   Effect of heeled ship with respect to different waterlines........   116
Fig. 14   Cross curves of stability   116
Fig. 15   Inversion of ship with large intact superstructure..............   117
Fig. 16   Cylinder rotated by force F.................................   117
Fig. 17   Kinetic energy of rotating particles   118
Fig. 18   Definition of dynamical stability   119
Fig. 19   Wind heeling moment and GZ variation with ship heel
           angle   120

Chapter 11
Fig. 1   Load waterline..   124
Fig. 2   Floodable length   125
Fig. 3   Immersed areas   126
Fig. 4   Area curve...   127
Fig. 5   Floodable length curve   127
Fig. 6   Flooded compartment   128
Fig. 7   Change of displacement at LCF   129
Fig. 8   Trapezium   131

Chapter 12
Fig. 1   Draught measurements.......................................   136
Fig. 2   Draught measurements.......................................   136
Fig. 3   Straight ways..   137
Fig. 4   Cambered ways   137
Fig. 5   Straight ways..   138
Fig. 6   Prior to Stern lift...   138
Fig. 7   Launching curves...   139

Chapter 13
Fig. 1   Ship length as stated in ICCL-66............................   146
Fig. 2   Damage stability requirements pre-2009   147
Fig. 3   Waterlines used in probabilistic damage calculations
           IMO2008c..   151
Fig. 4   Example of subdivision (IMO IMO2008c)   152
Fig. 5   Example 1 of how the subdivision length is determined   155
Fig. 6   Example 2 of how the subdivision length is determined   155
Fig. 7   Example 3 of how the subdivision length is determined   155
Fig. 8   Legend for Figs. 5, 6 and 7................................   155
Fig. 9   Examples of how the subdivision length is determined   158
Fig. 10   Effects of draught on GM (IMO IMO2008c)................   159

Chapter 14

Fig. 1   Multi-tiered structure of the second generation intact stability criteria 172

Fig. 2   Comparison of simplified waterline versus waterline of real wave in wave crest condition a 173

Fig. 3   Definition of the draft ith station with jth position of the wave crest.. 176

Fig. 4   Vulnerable ship stability curve 181

Chapter 15

Fig. 1   Examples 3 question 1 191

Fig. 2   Examples 3 question 4 192

Fig. 3   Examples 5 question 1 192

Fig. 4   Examples 3 question 2 193

Fig. 5   Examples 3 question 6 193

Fig. 6   Examples 2 5 question 5............................... 199

Fig. 7   Examples 5 question 6 199

List of Tables

Chapter 1
Table 1   Typical deadweight coefficients.   8
Table 2   Ship-type coefficients (Froude Style).   9
Table 3   Ship fineness coefficients   11

Chapter 2
Table 1   Example of calculation of Centre of Gravity using basic
           ship weights   22

Chapter 6
Table 1   Integration using Simpson's rule in tabular form.   62

Chapter 7
Table 1   Variation of hydrostatics with waterline   76
Table 2   Corrections to lightship displacement   79

Chapter 8
Table 1   Trial and error estimation for moulded beam   91

Chapter 9
Table 1   Estimation of heeled angle using Eq. (4)   103

Chapter 13
Table 1   Current damage stability regulations.   145
Table 2   IMO instruments containing deterministic damage stability   145
Table 3   Overview of damage stability conventions for different ship
           types.   149
Table 4   Parameters used in R index.   160
Table 5   Parameters used in p_i index   162
Table 6   Permeability regulations   169
Table 7   Permeability regulations for cargo ships   169

Chapter 14

Table 1   Wave cases for parametric rolling evaluation   175
Table 2   Corresponding encounter speed factor K_i   176
Table 3   Environmental conditions for pure loss.   179

Chapter 15

Table 1   Ship information   188
Table 2   Ship components weights   189
Table 3   Ship particulars   197
Table 4   Ship displacement versus draught   201
Table 5   Righting levers against heel angle.   202
Table 6   Ship angle of heel.   202
Table 7   Launching problem   203

Introduction to Naval Architecture

1

1 Introduction to Maritime Transportation

Transportation is an economic function serving along with other productive functions in the production of goods and services in the economy. If we can define production as the creation of utility, i.e. the quality of usefulness, then transportation creates the utility of place and time. That is to say goods may have little or no usefulness in one location at one time but may have great utility in another location at another time. One, naturally, has to bear in mind that some goods are so common as to be present almost everywhere and little can be gained by transporting them. Other goods may be unique and valuable so that they can be profitably transported great distances. Nevertheless as economic policies change, market barriers disappear and transportation costs reduce, even in the former case the benefits of transportation may outweigh the costs of producing locally.

The competition to sea transportation comes from road, rail, air and pipelines. As far as passenger transportation is concerned, there is a passenger/car ferry market benefiting from uniqueness of access (such as islands, or poor road/rail network), ease of access and competitive prices which manages to compete with air, road and rail transportation wherever these conditions prevail. In addition there is a lucrative cruise ship market. In the case of goods, transportation by sea dominates the inter-continental and international trade particularly in areas which are inaccessible by road and rail or where these networks are underdeveloped. Nevertheless, reduction of maritime transportation costs has been achieved at the expense of increase in size and the emergence of special types of ships which have special requirements for loading/unloading installations (e.g. deep water ports, special cranes for containers, special loading facilities for bulk carriers, tanker terminals, ramps for ro-ro ships). It may be a long while for the economics of large-scale transportation of goods to move in favour of the airborne trade. Department of Transport Statistics on the UK Seaborne trade (published annually by HMSO) clearly illustrate this point, although

© Springer International Publishing AG, part of Springer Nature 2018
P. A. Wilson, *Basic Naval Architecture*,
https://doi.org/10.1007/978-3-319-72805-6_1

this may be construed as an extreme example, UK being an island. Inland transportation by way of water is very small, with a few exceptions such as the Great Lakes and other inland seas and canal transportation. Pipelines over land and linking to offshore facilities are in competition with sea transportation of liquid and gaseous commodities.

The two most important recent developments in the field of maritime transportation are unitization and shipping in bulk. The former relates to the standardisation of dry cargo to improve its flow rate by means of palletization and containerization. The latter refers to the increase in size of vessels, either in the transportation of liquid or dry commodities, so that there can be benefit from reduced unit transportation costs. These moves contributed towards creating an integrated marine transportation system.

2   World Seaborne Trade

Seaborne trade is widely spread around the world. Nevertheless, the largest importers are the developed economies of North America, Western Europe and Japan. They account for approximately two thirds of the seaborne imports and, thus, have a dominant influence on seaborne trade. The developing countries comprising Central and South America, South-East Asia and Africa account for half the world's seaborne exports and a quarter of imports.

Such statistics do not always include information on the, then Soviet Union and the Eastern Block countries, and China whose share of the world seaborne trade is estimated at 6%.

Shipping is a complex industry and the conditions which govern its operations in one sector do not necessarily apply to another. It might even, for some purposes, be better regarded as a group of related industries. Its main assets, the ships themselves, vary widely in size and type; they provide the whole range of services for a variety of goods whether over shorter or longer distances. Although one can, for analytical purposes, usefully isolate sectors of the industry providing particular type of service, there is usually some interchange which cannot be ignored. Most of the industry's business is concerned with international trade and inevitably it operates within a complicated world pattern of agreements between shipping companies, understandings with shippers and policies of governments.

The above statements illustrate succinctly the complexities of shipping economics which in turn affect a wide range of directly and indirectly involved industries. When one examines the types of cargo carried by sea and their share in the seaborne trade, the indications are that crude oil and the so-called five major dry bulk commodities dominate the seaborne trade. These are, furthermore, confirmed by statistics relating to the world fleet tonnage and the new orders placed (see, for example, Classification Society Annual Reports).

It is convenient to classify the seaborne trade in terms of *parcels* where a parcel is an individual consignment of cargo for shipment. It is, therefore, possible to classify world shipping in two broad-based categories, namely bulk shipping for transportation of dry and liquid bulk cargo and liner shipping for general cargo in various forms and shapes. The former carries big parcels filling a whole of a ship or a hold, whilst the latter carries small parcels which need to be grouped together with others for transportation. Liner shipping services are, usually, regular, i.e. at given times between specified ports. Different types of ships can, in general, be assigned to one or other of these two categories. One interesting feature is that shipping (or even production) companies need not own their ships but hire them. This is called chartering and varies from voyage charter from one end of the spectrum to the bareboat charter at the other end where the ship owner is not involved in the operation of the vessel.

For more information on these subjects consult: Maritime Economics by M. Stopford (published by Unwin Hyman).

The diversity of the vessels classified according to the function they perform and the type of cargo they transport (i.e. their mission)—size in excess of 100 GRT—can be found in a paper entitled: *Matching merchant ship designs to markets* by I.L. Buxton (published in the Transactions of the North East Coast Institution of Engineers and Shipbuilders - Vol. 98, pp. 91–104, 1982). Although this table dates to 1979, it gives a good impression of the state of the world fleet. Note GRT: 1 Gross Register Ton = 100 cubic feet; this is a measure of volume of spaces below tonnage deck and between deck spaces above tonnage deck and all permanently closed spaces above upper deck.

On the other end of the spectrum, we have the vessels whose mission is not transportation but provision of specialised services and support and the performance of a special function. In this category of vessels one can include fishing vessels, tugs, dredgers, drilling ships, pipe-laying ships, cable-laying ships, survey vessels (such as oceanographic research, hydrographic survey, seismic exploration ships), supply vessels, diving support vessels, fire service boats, life boats, submersibles, a large range of naval vessels which are becoming very specialised (e.g. the Single Role Mine Hunter) and a large variety of small craft which are mainly used for leisure purposes.

In addition, there are various structures fixed to the seabed (jackets and jack-up rigs) or tethered (various types of semi-submersibles), used in the exploration and production of offshore oil and gas, providing a stable platform for operations in severe weather conditions. In this category, we can also include various offshore loading towers.

3   Survey of Maritime Vessels

The Means of Support

In the previous section, maritime vessels were classified according to the function they perform, i.e. their mission. This classification, however, does not provide information on the form of support of the vessel during its operation. For example in the case of ferries the service can be provided by a mono-hull, a catamaran, a Small Water-plane Area Twin Hull (SWATH), a hovercraft, a hydrofoil etc. The basic concept of a vessel being a single shell floating according to Archimedes' principle, containing within it sufficient elements ensuring its integrity and providing the capability to perform its mission successfully has been challenged, in particular in the twenty-first century, in order to improve feasibility and efficiency. Parallel to improvements in the means of propulsion of the displacement hulls, other types emerged based on challenging or improving the conventional displacement hull. In this respect the classification shown in the book: Modern Ship Design by T.C. Gillmer (published by the Naval Institute Press, USA, 1984) according to mode of support whilst operating in or on the sea surface is very useful. It provides a good summary of the various vessel forms available for the execution of a particular mission. Thus, we have:

Aerostatic support: These are basically air buoyant structures as the self-induced low-pressure air cushion provides an upward force lifting all or most of the hull from water and, thus, eliminating all or most of the drag associated with motion through the water. Typical examples in this category are: the hovercraft, which glides slightly above the water surface and is amphibious; the side wall hovercraft, usually referred to as a Surface Effect Ship (SES), where the air cushion builds up under the rigid hull sides, rather than the skirt as in the previous case—some contact with the water is still maintained and, thus, this type is not amphibious.

Hydrodynamic support: These are based on the provision of dynamic support produced as a result of rapid forward motion. In the case of hydrofoils (surface piercing or submerged) as the foil cuts through the water at speed considerable lift is generated, thus supporting the hull above the water on legs attached to the foils. In the case of planing hulls lift is generated by the shallow V form of the hull. This is a less efficient means of hydrodynamic support and usually restricted in size due to power requirements and induced structural stresses and also restricted to operating in reasonably calm weather. In addition, there is the semi-planning or semi-displacement hull which combines a reasonably high-speed and rough water performance.

Hydrostatic support: These are vessels which float on the surface of the water with the buoyancy equal to their weight. This category includes vessels of wide-ranging sizes, from the very large and deep-draught vessels, such as Very Large Crude Carrier (VLCC), to small coastal vessels. The underwater shape of the hulls varies considerably depending on mission requirements (e.g. high forward speed). Multi-hulled vessels are also included in this category. In the case of the Small Waterplane Area Twin Hull (SWATH) the buoyancy is mostly provided by the pontoons placed well below the free surface and struts with narrow waterline support the spacious deck structure. They posses good wavemaking resistance properties and good seakeeping performance thus providing a stable platform for operations. Other multi-hulls, namely catamarans and trimarans, also provide large working spaces above the water and stable platforms for operations. Finally, the submersibles (big submersibles are usually referred to as submarines) are a special case of this category, as they are vessels which can operate partly or totally immersed and abide by Archimedes' principle.

4   A Global View of Ship Design—the Design Requirements

A ship is designed for a purpose. This may be:

1. To give pleasure to a yachtsman
2. To deploy a weapons system
3. To carry cargo or passengers; to provide a service.

To fulfil this purpose the ship must:

1. Have enough internal capacity to contain everything requiring to be stowed in the ship.
2. Be divided internally into compartments serving a specific function (e.g. machinery space, accommodation, cargo holds). Each compartment must be of a size suitable for its function, must be positioned suitably within the ship in relation to other compartments (e.g. a galley adjacent to a dining saloon) and accessible via appropriate passage and stairways. Each compartment must be suitably equipped.
3. Float at its designed waterline when fully loaded, and float reasonably level. This is important from the point of view of seaworthiness and manoeuvrability. Excessive trim bow or stern down will make steering difficult and may result in excessive amounts of water coming on deck in rough weather. The draught to which a merchant ship may be loaded is governed by law.
4. The ship must be stable and float upright in calm water. It should also be safe from capsize in rough weather. The ship should be stable in all conditions of loading and should be capable of sustaining a reasonable degree of damage (resulting in partially flooding the hull) without sinking or becoming unstable.
5. The hull must be so shaped that it does not require an excessive amount of propulsion power to achieve its service speed.
6. The hull must be so shaped that it does not pitch, roll or heave excessively in rough weather, and so that it does not get excessive amounts of water on deck or experience slamming damage.
7. The hull structure must be strong enough to sustain the loads applied to it in service. The structure must not vibrate excessively. The structure should not deteriorate too rapidly in service (e.g. through corrosion).
8. The power installed in the ship must be adequate for the required service speed and there must be enough fuel capacity for the required operating range.
9. The vessel must represent value for money, i.e. be so designed as to maximise return on capital invested.

---

5   The Design Cycle

It is typical of a design calculation that it is necessary to know the result of the calculation before the calculation can be carried out.

Consider the structural design problem:

- The structure must be strong enough to sustain the loads applied to it.
- A substantial part of the loading is associated with structural weight.
- The structural weight depends on the size of the structural components.
- The size of the components depends on the strength required.

Thus to estimate the required structural strength you need to know the structural weight which cannot be found unless you already know how strong the structure needs to be.

Likewise consider the power problem:

- The power required to propel the ship depends on its all up weight (amongst other considerations).
- Major items of weight are propulsion machinery and fuel.
- These items of weight cannot be estimated until the power to propel the ship is known.

In order to proceed with the technical design process in this sort of circumstance the starting point has to be a good first guess at the answer. A typical procedure would be

1. Guess total weight for ship and contents.
2. Carry out a power calculation, select the size and type of main engine, estimate machinery and fuel weights.
3. Carry out a structural calculation, select the sizes of structural components and estimate structural weights.
4. Tot up the total weight including structure, machinery, fuel, cargo, equipment and stores.
5. Compare new total weight with the starting guess and repeat the calculation if the discrepancy is too large.

It is not usual to exactly repeat the sequence of calculations. Initially very simple methods of estimation that yield quick approximate answers will be used. Subsequently more rigorous, but more time-consuming methods will be used. The final stages of confirming the design may well involve the use of expensive computer programmes or expensive model testing.

6   The Geometry of a Ship's Hull

One of the earliest design tasks for the Naval Architect is to define the shape of the outer surface of the ship's hull. Everything carried on board below the upper deck has to fit inside this surface and the shape chosen must be seaworthy and economic to propel.

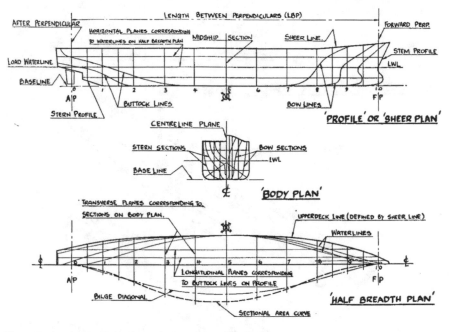

Fig. 1   Body plan (with kind permission of Dr. John Wellicome)

The shape of the hull is defined on a Linesplan or Sheer Drawing, which is, in effect, a contour map of the hull surface (see Fig. 1).

Three views are shown:

1. A Profile or Sheer Plan which is a side view of the hull.
2. A Half-Breadth Plan which is a view from above.
3. A Body Plan which is a view from in front or behind.

Most common terms used to describe a ship's geometry are also shown in Fig. 1. Lines drawn on these three views mostly present lines of intersection between the hull moulded surface and various transverse, longitudinal, horizontal or diagonal planes. Some lines such as the line of the upper deck at side will not lie in any plane, but the projections of such lines will be shown in all three views. In merchant shipbuilding moulded dimensions are to the outer edge of the ship frames. The shell plating lies outside the moulded form. Small craft and Warship practice is to take the outer surface of the plating as the moulded surface.

Intersections with horizontal planes are generally called **waterlines** but may be called **level lines** above the load waterline.

Intersections with longitudinal planes are generally called **buttock lines**, but may be called **bow lines** ahead of amidships (Table 1).

Intersections with vertical planes are generally called **stations** or **sections**. These are illustrated in Fig. 2.

Table 1 Typical deadweight
coefficients

Supertanker	$C_D = 0.78$
Container ship	$C_D = 0.56$
Hydrofoil ferry	$C_D = 0.30$

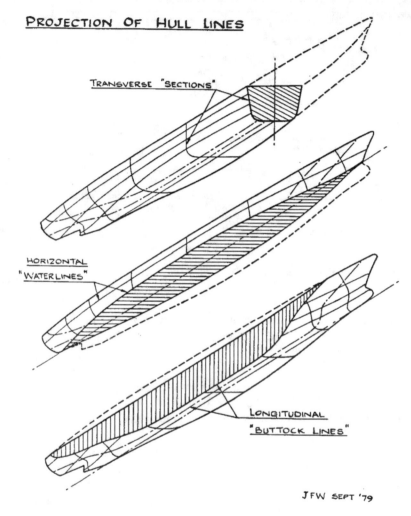

PROJECTION OF HULL LINES

TRANSVERSE "SECTIONS"

HORIZONTAL
"WATERLINES"

LONGITUDINAL
"BUTTOCK LINES"

JFW SEPT '79

Fig. 2 Projection of hull lines (with kind permission of Dr. john Wellicome)

7   Comparative Design Parameters

The starting guess is usually obtained by comparing existing ships to the new design.
Some typical ship parameters used for comparison are:

7.1 Deadweight Coefficient

$$C_D = \frac{\text{Load that can be carried}}{\text{All up weight}} = \frac{\text{Deadweight}}{\text{Displacement weight}}$$

Deadweight in this context comprises all those items that are not part of the fabric of the ship, i.e. cargo + fuel + stores + ballast + etc. Please note that Displacement Weight = Lightweight + Deadweight, where Lightweight comprises all those items that are part of the fabric of the ship, i.e. structure, machinery, outfits, superstructure.

Typical values:

Deadweight coefficient is frequently used to obtain a first estimate of the total weight or displacement weight corresponding to a given cargo-carrying capacity.

7.2 Slenderness Coefficients

There are several alternative parameters used to express the relation between displacement and hull length. The most frequently found are:

$$\text{Taylor Displacement-Length Ratio} = \frac{\text{Displaced Mass (tons)}}{[\text{Length (ft)}/100\,]^3} = \frac{\Delta}{\left(\frac{L}{100}\right)^3}$$

$$\text{The ITTC Volumetric Coefficient } C_V = \frac{\text{Displacement volume}}{[\text{Length}]^3} = \frac{\nabla}{L^3}$$

$$\text{Froude Displacement Coefficient } @ = \frac{\text{Length}}{[\text{Displacement volume}]^{1/3}} = \frac{L}{\nabla^{1/3}}$$

The last two coefficients are non-dimensional and should be preferred. Typical values of these slenderness coefficients are given in Table 2.

Table 2 Ship-type coefficients (Froude Style)

Ship type	$@$	$10^3 C_V$	$\frac{\Delta}{\left(\frac{L}{100}\right)^3}$
Racing VIII	17.00	0.2	6
Frigate/Destroyer	7.5	2.5	70
Light displacement racing yacht	7.0	3.0	80
Container ship	6.5	3.5	105
Large tanker	5.0	8.0	230
Cruising yacht	4.75	9.5	270
Salvage tug	4.25	13.0	370
Seventeenth-century first rate	4.00	15.5	450

7.3 Fineness Coefficients

Some hulls (e.g. Dutch Sailing Barges) have very full rounded ends, others (e.g. frigates) have very knife-like ends. Here we are describing not the relation between the displacement and length but some characteristic of shape. Various coefficients are used to characterise this shape and are collectively known as fineness coefficients. The commonly used fineness coefficients are:-

$$\text{Block Coefficient} = \frac{\text{Volume of underwater form}}{\text{Length} \times \text{Beam} \times \text{Draught}} = \frac{\nabla}{L\ B\ T} = C_B$$

$$\text{Prismatic Coefficient} = \frac{\text{Volume of underwater form}}{\text{Length} \times \text{Max Cross Sec. Area}} = \frac{\nabla}{L A_m} = C_P$$

$$\text{Maximum Sectional Area Coefficient} = \frac{\text{Max Cross Sec. Area}}{\text{Beam} \times \text{Draught}} = \frac{A_m}{B\ T} = C_M$$

Note that usually in the big ship world at least the maximum section is at the midship section, halfway along the hull (Station 5 on a normal lines plan). C_M is then simply the amidships sectional area coefficient.

Note:

$$C_B = \frac{\nabla}{L\ B\ T} = \frac{\nabla}{L\ A_m}\frac{A_m}{B\ T} = C_P\ C_M$$

i.e.

$$C_B = C_P C_M$$

thus these three coefficients are not independent.

$$\text{Waterplane Area Coefficient} = \frac{\text{Waterplane Area}}{\text{Length} \times \text{Beam}} = \frac{A_w}{L\ B} = C_W$$

There is no simple relation between C_W and C_B or C_P, but the shape of the hull sections near the ends is governed by this relationship.

Increasing C_W without altering the sectional area curve (and hence without changing C_P) will produce more V-shaped end sections (see Fig. 3).

The fineness coefficients appropriate to a particular vessel depend on the speed she is designed for and on seakeeping considerations. Without going into the technical reasons at this stage, it is true to say that faster ship types need a finer hull form if they are to achieve their speed economically in terms of propulsive power requirement. Finer forms also have easier motions and are subject to less slamming in rough weather. Typical Fineness Parameters are given in Table 3.

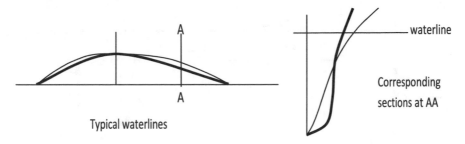

Fig. 3 Waterlines

Table 3 Ship fineness coefficients

Ship type	C_p	C_M	C_B	C_W
Trawler	0.648	0.880	0.570	0.720
Car ferry	0.551	0.920	0.507	0.640
Fast cargo liner	0.664	0.980	0.650	0.749
Cargo tramp.	0.735	0.980	0.720	0.803
Tanker	0.842	0.985	0.830	0.887
Sailing yacht excluding fin keel	0.550	0.680	0.374	0.700

7.4 Speed Parameters

Ship speed is judged in relation to ship size. The parameters used are,
Froude Number.

$$F_n = \frac{V}{\sqrt{gL}}$$

where V (m/s), L (m) and g (m/s^2) all are in consistent units (Fig. 4).
and,

$$\textbf{Taylor Speed-Length Ratio} = \frac{V_k}{\sqrt{L}}$$

where V_k (knots) and L (ft).

7.5 Design Trend Lines

For initial estimating purposes it is common practice to plot various ship parameters against speed to establish trends in parameter variation. Data from a variety of commercial and naval ship types is shown in Figs. 5 and 6.

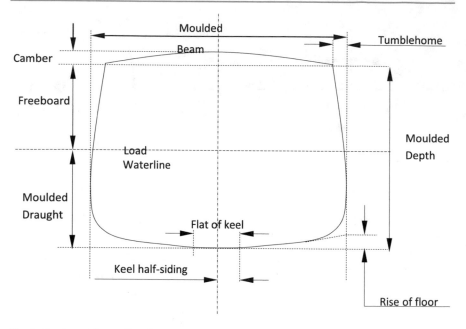

Fig. 4 Section and waterline shapes

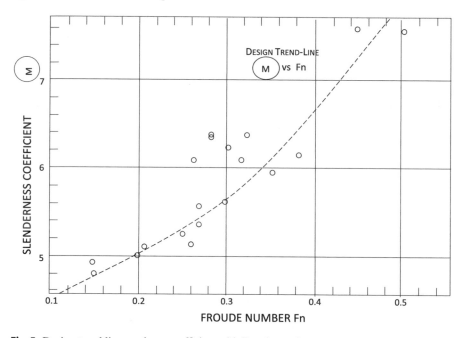

Fig. 5 Design trend lines, volume coefficient with Froude number

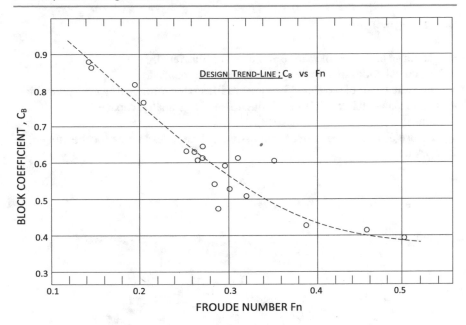

Fig. 6 Design trend line block coefficient variation with Froude number

7.6 Displacement Mass and Weight

The normal notation and units are as follows: ∇ = displacement volume m^3

Δ = displacement mass (tonnes or kg in small boats)

or Δ = displacement weight (kN or MN)

Note the symbol is used for both mass and weight—the units used must be stated to make clear which is intended

$$\Delta\text{mass} = \rho\nabla$$

$$\Delta\text{weight} = \rho g\nabla = \text{Buoyancy (equilibrium case)}$$

ρ = water density: Freshwater $\rho_F = 1000\,\text{kg/m}^3 = 1.000\,\text{tonne/m}^3$
Standard Saltwater $\rho_s = 1025\,\text{kg/m}^3 = 1.025\,\text{tonne/m}^3$

Saltwater density is a function of salinity primarily, although pressure and temperature do have a slight effect.

8 Terms Used to Define the Midship Section

The main terms used to define the midship section are illustrated in Fig. 4.

9 Summary

- The range and types of marine vehicles are illustrated.
- The complexities of ship design process are introduced.
- The basic terms of naval architecture are illustrated through a typical Lines Plan.
- The fundamental coefficients relating to ship's geometry, capacity and speed are defined.
- The use of design trend lines is illustrated through examples for the student to work on.

Basic Properties

1 Mass, Weight and Moments of Weight

You should already be familiar with the following concepts:

MASS: A measure of the amount of material in a body. In SI units mass is measured in kilograms (kg) or in our context in tonnes (1 tonne $= 1000$ kg).

WEIGHT: The vertical force acting downwards on a given mass in a gravitational field. This force is dependent on the body mass and the local acceleration due to gravity:

$$w = mg$$

where $w =$ weight measured in newtons (N), kilonewtons (kN) or meganewtons (MN) and $g = 9.81$ m/s^2 as a good average earth value of gravitational acceleration. Note: 1 N $= 1$ kg 1 m/s^2 1 kN $= 1000$ N, 1 MN $= 1000$ kN $= 10^6$ N. The weight of a 1 tonne mass is 9.81 kN.

2 Moment of Weight

The moment of a force about a point is the product of the force and the perpendicular distance between the point and the line of action of the force. In the case shown in Fig. 1

$$M = wx.$$

Moments are measured in N m, kN m or MN m, i.e. newton metres, kilonewton metres, etc. For a system of masses, held in a frame, to balance about a particular pivot point P the algebraic sum of all the moments of weight about P must be zero.

© Springer International Publishing AG, part of Springer Nature 2018
P. A. Wilson, *Basic Naval Architecture*,
https://doi.org/10.1007/978-3-319-72805-6_2

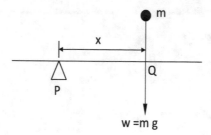

Fig. 1 Definition of moment

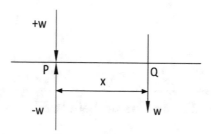

Fig. 2 Addition of equal forces at a singular point

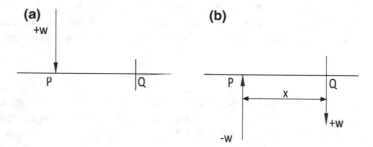

Fig. 3 Forces replaced by a couple plus force

3 Transfer of Weight—Equivalent Forces and Weights

Introducing equal and opposite forces $+w$ and $-w$ sharing the same line of action is in effect to leave the force system unchanged (see Fig. 2).

The new force system is equivalent to the sum of two separate force systems (see Fig. 3).

(a) is a **force** w transferred from Q to P
(b) is a **couple** (i.e. a pure moment with no net force).

Fig. 4 Moment of a couple

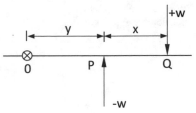

Fig. 5 Multiple forces added
for equivalent moment on
system

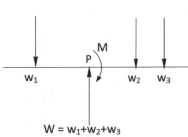

Thus the original force acting through Q is equivalent to an equal force acting
through P plus a couple $w \times x$. The moment of the couple about 0 is (see Fig. 4):
$M = +w(x + y) - wy = wx$.

That is, the moment is independent of y and hence of the position of 0.

Thus **couple** is simply a moment and not a moment about any particular point.

4 Centres of Gravity

By introducing an appropriate couple, which would cause the system to rotate if it
is nonzero, the forces of weight of a system of masses can be transferred to a given
point P and aggregated into a single force equal to the sum of all the weights of the
masses in the system (see Fig. 5).

If the point P is chosen so that the net couple is zero, then the system would
balance on a pivot at P.

**If the system can be turned round to any required attitude and still remains
balanced about P, then P is called the** *Centre of Gravity* **of the mass system.**

5 Summation Notation

Consider a set of masses $m_1, m_2 \ldots$ at points $x_1, x_2 \ldots$ on a horizontal beam. Let N
be the total number of masses (see Fig. 6)

The weights w_i of these masses m_i can be transferred to a pivot point P to give
an equivalent single force:

Fig. 6 Generalisation to
large number of point masses

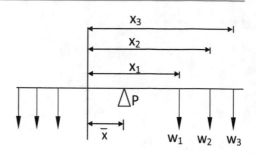

$$W = w_1 + w_2 + w_3 + \cdots + w_N = \sum_{i=1}^{N} w_i \tag{1}$$

together with a couple,

$$M = w_1(x_1 - \overline{x}) + w_2(x_2 - \overline{x}) + w_3(x_3 - \overline{x}) + \cdots + w_N(x_N - \overline{x})$$

or

$$M = \sum_{i=1}^{N} w_i(x_i - \overline{x}) \tag{2}$$

In the above notation w_i is a member of the set $w_1, w_2 \ldots w_N$ and x_i of the set $x_1, x_2, x_3 \ldots x_N$ and the notation $\sum_{i=1}^{N}$ implies the summation of all terms of the indicated type [w_i or $w_i(x_i - \overline{x})$ as appropriate], summed for all values of the subscript i between $i = 1$ and $i = N$.

6 Estimation of Point of Balance

The expression (2) for the moment M can be rewritten as:

$$M = \sum_{i=1}^{N} w_i x_i - \overline{x} \sum_{i=1}^{N} w_i$$

or, on making use of (1),

$$M = \sum_{i=1}^{N} w_i x_i - W.\overline{x}.$$

Fig. 7 2-D rotation of axis
system

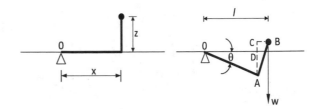

The system of masses balances at P if $\bar{x}$ is chosen to give,

$$M = 0.$$

That is, if,

$$\bar{x} = \frac{\sum\limits_{i=1}^{N} w_i \, x_i}{W} \tag{3}$$

7 The Effect of Rotation on Moment Acting

After rotating clockwise through an angle θ a mass m (of weight w), whose coordinates relative to the pivot 0 are initially (x, z), will exert a weight moment $w \times l$ (see Fig. 7).

Geometrically,

$$l = OD + CB = x \, \cos\theta + z \, \sin\theta \tag{4}$$

Thus, the moment is

$$M(\theta) = w \, x \, \cos\theta + w \, z \, \sin\theta \tag{5}$$

In other words the **vertical moment** $w \times z$ is as important as the **horizontal moment** $w \times x$ in deciding turning moments once a body is rotated from its initial position.

8 General Expressions for Centre of Gravity

Consider an axis system $Oxyz$ obtained by rotating axis system $OXYZ$ about axis Oy through an angle θ. Consider a set of masses of weight $w_i (i = 1, 2, \dots N)$ fixed at points (x_i, y_i, z_i) defined with reference to axis system $Oxyz$ (see Fig. 8).

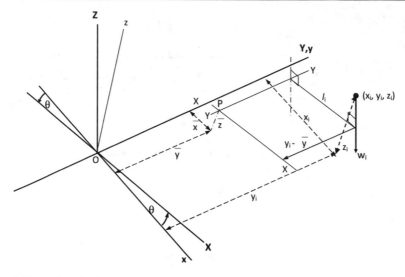

Fig. 8 3-D rotation of axis system

Imagine the system balanced on a point P whose coordinates, in the $Oxyz$ axis system, are $(\bar{x}, \bar{y}, \bar{z})$.

If the system is balanced, that is if P coincides with the **Centre of Gravity** of the set of masses, then moments of weight about the horizontal axes PX and PY must both be zero. As discussed earlier, after rotation the lever about the axis PY, according to Eq. 4,

$$l_i = (x_i - \bar{x})cos\theta + (z_i - \bar{z})sin\theta$$

so that the moment about this axis is

$$M_{YY} = cos\theta \sum_{i=1}^{N} w_i(x_i - \bar{x}) + sin\theta \sum_{i=1}^{N} w_i(z_i - \bar{z}) \qquad (6)$$

The moment about the axis PX is

$$M_{XX} = \sum_{i=1}^{N} w_i(y_i - \bar{y}) \qquad (7)$$

Both M_{YY} and M_{XX} are zero for all rotations if

$$\sum_{i=1}^{N} w_i(x_i - \bar{x}) = 0, \quad \sum_{i=1}^{N} w_i(y_i - \bar{y}) = 0, \quad \sum_{i=1}^{N} w_i(z_i - \bar{z}) = 0 \qquad (8)$$

On writing

$$W = \sum_{i=1}^{N} w_i$$

as before, these three equations are satisfied if

$$\overline{x} = \frac{\sum_{i=1}^{N} w_i \, x_i}{W}, \quad \overline{y} = \frac{\sum_{i=1}^{N} w_i \, y_i}{W}, \quad \overline{z} = \frac{\sum_{i=1}^{N} w_i \, z_i}{W} \tag{9}$$

The point $P = (\overline{x}, \overline{y}, \overline{z})$ is the **Centre of Gravity** of the system of masses.

Although these expressions (9) involve **weight** rather than **mass**, both numerators and denominators contain factors g (since $w_i = m_i g$) and it is equally possible to write, in terms of mass:

$$\overline{x} = \frac{\sum_{i=1}^{N} m_i \, x_i}{M}, \quad \overline{y} = \frac{\sum_{i=1}^{N} m_i \, y_i}{M}, \quad \overline{z} = \frac{\sum_{i=1}^{N} m_i \, z_i}{M} \tag{10}$$

where

$$M = \sum_{i=1}^{N} m_i \,.$$

9 Example Calculations of Centre of Gravity

The following calculations (see Table 1) of longitudinal position (LCG) and vertical position (VCG) of the Centre of Gravity of a vessel, whose Centre of Gravity is presumed to be on centreline (i.e. $\overline{y} = 0$), illustrate the application of the above equations: Measurements x_i are longitudinal from amidships (positive forward), z_i vertically above the keel line.

Total ship mass $= 29465$ tonnes.

$LCG = +\frac{113570}{29465} = +3.85\,\text{m}$ (forward from amidships).

$VCG = \frac{213472}{29465} = 7.24\,\text{m}$ (above keel).

Table 1 Example of calculation of Centre of Gravity using basic ship weights

Item	m_i (tonnes)	x_i (m)	z_i (m)	$m_i x_i$ (tonne-m)	$m_i z_i$ (tonne-m)
Hull structure	6820	−8.0	8.7	−54560	59334
Main engines	1960	−55.0	6.8	−107800	13328
Anchors and cable	150	+95.0	11.5	+14250	1725
Lifeboats	10	−57.0	23.0	−570	230
Cargo	18450	+15.0	7.4	+276750	136530
Fuel	2000	−5.0	0.6	−10000	1200
Stores	75	−60.0	15.0	−4500	1250
	Σm_i			$\Sigma m_i x_i$	$\Sigma m_i z_i$
Total	29465			+113570	213472

10   Summary

- The concepts of (first) moment of a force and couple are summarised.
- The coordinates of the Centre of Gravity of a three-dimensional system of weights are derived.
- A typical calculation of a ship's LCG and VCG is illustrated.

Equilibrium and Stability Concepts for Floating Bodies

1 Pressures in a Uniform Incompressible Fluid at Rest

Consider a column of fluid ending at a free surface at atmospheric pressure p_o. Assume vertical walls and a uniform cross section of area A (see Fig. 1).

Net vertical pressure load on column = $(p - p_o)A$ upwards. Weight of fluid = $\rho g A h$ vertically downwards.

Equilibrium requires no net force, i.e. $(p - p_o)A = \rho g A h$

or

$$p - p_o = \rho g h .\tag{1}$$

Note: p is constant if h is constant, i.e. over any horizontal plane.

2 Pressures on a Closed Surface S at Rest in a Fluid also at Rest

A surface S immersed in the fluid can be considered as covered by a set of surface elements as shown in Fig. 2. An element of area S, as shown, will be subject to pressure p given by Eq. (1). It will have a net force **normal** to the surface due to p given by $p \delta S$ (if δS is sufficiently small).

The pressure loads can be summed vectorially to yield a net force on S. Taking moments of elementary forces about a suitable system of axes and summing to obtain net moments of force about the axes can identify the line of action of this force. The force will not depend on whether the surface S encloses some foreign body (e.g. a ship) or merely some more fluid. This is because it depends on pressures in the fluid outside S.

© Springer International Publishing AG, part of Springer Nature 2018
P. A. Wilson, *Basic Naval Architecture*,
https://doi.org/10.1007/978-3-319-72805-6_3

Fig. 1 Pressure forces on the
base of a vertical cylinder

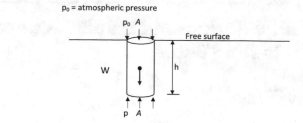

Fig. 2 Pressure forces on a
three dimensional body

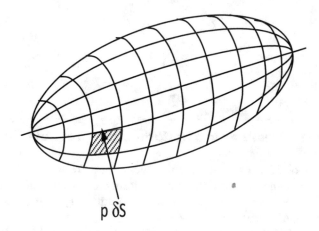

$$p \, \delta S$$

3 Archimedes Principle

Consider a surface S, enclosing a volume V, containing fluid at rest under:

1. The resultant hydrostatic pressure load evaluated above,
2. The weight of the fluid, acting vertically down through the centroid of the volume.

Since the fluid in S is in equilibrium under these two forces they must be self cancelling. Thus, the resultant pressure force on S must:

1. Be equal in magnitude to the weight of fluid in S
2. Act vertically upwards through the Centre of Gravity or centroid of the enclosed volume.

If the fluid in S is displaced and replaced by a body whose outer shell fits S then that body will experience the same net pressure force which will at least partially support the weight of the body itself.

This is the **Archimedes's principle**:

A body immersed, or partly immersed, in a fluid at rest experiences a buoyancy force which,

1. Has a magnitude equal to the weight of liquid displaced,
2. Acts vertically upwards through the centroid of the immersed volume of the body (henceforth called the **centre of buoyancy**.

---

4 Calculating Force of Buoyancy

In practice the force of buoyancy and centre of buoyancy can be calculated in one of two ways:

1. A calculation of loads from properties of the immersed volume using Archimedes principle.
2. A surface integration of hydrostatic pressure loads.

1. Is the traditional **ship** procedure,
2. As advantages for the complex geometries of offshore platforms.

---

5 Equilibrium and Stability of Floating Bodies

For any ship or offshore structure, or indeed any marine artefact which has (at some stage in its life) to float freely, it is necessary to ensure that:

1. The body floats freely in equilibrium in its design attitude;
2. The body is adequately stable in its equilibrium condition.

These two concepts are very different. Equilibrium is obtained when the forces of weight and buoyancy cancel precisely. The requirements for this are that:

1. The forces of weight and buoyancy are equal in magnitude,
2. The Centre of Gravity (of the body mass) is in line vertically with the centre of buoyancy.

5.1 General Definitions

Stability is to do with the body behaviour **after it has been disturbed from its equilibrium state**. This problem is best illustrated by considering a simple cone standing on a table (see Fig. 3) A cone standing precisely upright on its vertex with its Centre of Gravity vertically above the vertex is in equilibrium since weight and the reaction force at the table cancel out. However, if it is tilted, even only very slightly, the weight and reaction forces will form a couple that cause the cone to fall over onto its side (see Fig. 3). A cone standing on its base is likewise in equilibrium.

Fig. 3 Stability conditions
and types

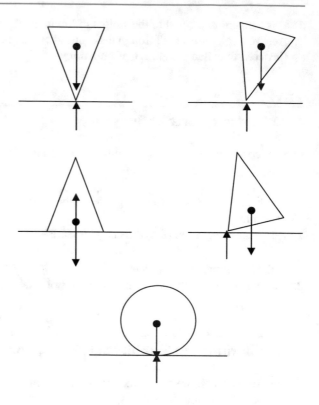

In this case the couple acting on a tilted cone is such as to cause it to sit back down on its base, provided the angle of tilt is not too large (see Fig. 3 middle).

A cone lying on its side is also in equilibrium. In this case if the cone is moved (by rolling to a new position) it will stay where it is left: neither tending to move back to where it came from nor to move further away from its initial position (see Fig. 3 bottom).

Stable Equilibrium: A body which, following some small disturbance, will tend to return to its equilibrium position on being released is in a condition of **STABLE** equilibrium.

Unstable Equilibrium: A body which, following some small disturbance, will tend to move further away from its equilibrium position on being released is in a condition of **UNSTABLE** equilibrium.

Neutral Equilibrium: A body which, following some small disturbance, tends neither to return to, nor move further away from its equilibrium position is said to be in a condition of **NEUTRAL** equilibrium.

Floating bodies will have conditions of stable or unstable equilibrium. If, like a submarine, the body is fully below the free surface it may also be neutrally stable. For reasons to be discussed later, a body floating in the free surface is most unlikely to be neutrally stable.

Fig. 4 Stability of a fully
submerged body

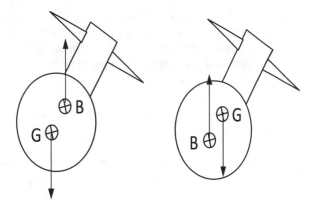

5.2 Stability of a Submerged Body

For a submerged body the position of the centre of buoyancy B is fixed by the shape
of the buoyant volume: its Centre of Gravity G by the distribution of mass within the
body. The body can be in **equilibrium** if the magnitudes of weight and buoyancy are
equal and B and G are vertically in line, as shown in Fig. 4. Clearly the equilibrium
is stable if G is below B (see Fig. 4 left), unstable if G is above B (see Fig. 4 right)
and neutral if B and G coincide.

5.3 Stability of a Floating Body

The stability of a body floating in the free surface is more complicated than the
submerged body, because the shape of the displacement volume changes and, hence,
the position of the centre of buoyancy changes, as the body is rotated from its equilibrium position. The vertical position of B is not now important and B may in fact
be below G. The body is stable if the point M on the diagram is above G (see Fig. 5)

Fig. 5 Effects of positions of
centre of buoyancy and
gravity

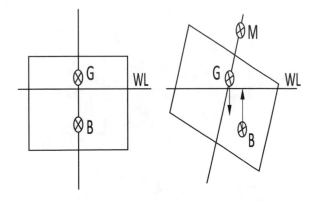

and unstable if M is below G. Because the point M is itself not fixed, but rises as the angle of rotation increases, the neutral equilibrium condition (M and G coincident for all small rotations) is actually not attainable. The lowest position of M is called the **metacentre** of the floating body–more on this later!

6 Summary

- Hydrostatic pressure and Archimedes's principle are summarised.
- Equilibrium conditions for weight and Buoyancy forces are introduced.
- The general concept of stability is defined, and stable, unstable and neutral equilibrium are illustrated.
- The concept of stability is illustrated for floating and submerged vessels.

Calculating Volumes and Centres of Buoyancy

4

In order to study the properties of a floating body, such as a ship, it is necessary to be able to calculate displacement volume and centre of buoyancy. To calculate weights and centres of gravity of a deck plate or a bulkhead plate it is necessary to calculate the area and centre of area of a plate whose outline is a curve defined by the hull shape. We need to be able to calculate areas and centres of area of a uniform plane lamina or the volumes and centres of volume of a uniform three-dimensional solid. The second process (finding volume properties) is an extension of the first and both involve processes of integration.

The volume of a small slice across the ship is (see Fig. 1):

$$\delta \nabla = a(x)\delta x$$

where $a(x)$ is the cross-sectional area of the underwater hull form for this slice of the ship. The volume of the ship is found by calculating $a(x)$ for a large number of points along the ship length and adding up the volumes of each slice of the ship. This is equivalent to plotting the curve showing how $a(x)$ varies along the ship length and then calculating the area under this curve (see Fig. 1). Thus calculating volume properties can be reduced to calculating the properties of an appropriate lamina. Moments of volume can be found by evaluating moments of area of the sectional area curve.

1 Integration as a Limit of Summation

Integration can be regarded in two ways:

1. As the inverse of differentiation and
2. As the limit of a summation process.

© Springer International Publishing AG, part of Springer Nature 2018
P. A. Wilson, *Basic Naval Architecture*,
https://doi.org/10.1007/978-3-319-72805-6_4

Fig. 1 Sectional area curve

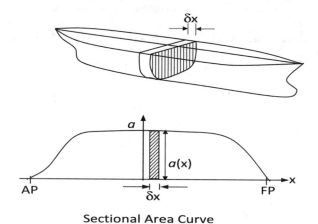

Sectional Area Curve

Fig. 2 Integration as limit of summation

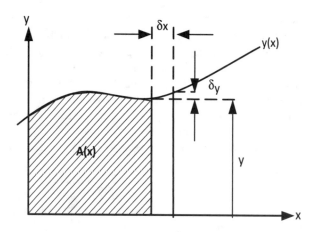

Analytically (1) yields the most useful approach to standard integration formulae, but in terms of modelling physical systems in engineering the second approach (2) is of immense practical use. Consider a function $A(x)$ representing the area under a curve $y(x)$ (see Fig. 2). As x increases by a small amount, δx the area increases by an approximately rectangular strip of area,

$$\delta A \approx y(x)\delta x.$$

In fact,

$$y(x)\delta x \leq \delta A \leq (y + \delta y)\delta x$$

as drawn. Thus as

$$\delta x \to 0,$$

we have

$$\frac{dA}{dx} = y(x).$$

This yields $A(x)$, by the inversion route (1) as

$$A(x) = \int y(x)dx.$$

As an alternative view consider the area under $y(x)$ between x_1 and x_N to be divided into strips of width, δx_i (see Fig. 3).

Neglecting the small-shaded parts the area is approximately,

$$A = \sum_{i=1}^{N} y(x_i)\delta x_i = \sum_{i=1}^{N} y_i\,\delta x_i$$

if $y(x_i)$ is simply written as y_i. The number of strips can now be increased many fold so that $N \to \infty$ and $\delta x_i \to 0$.

Clearly, as this happens, the shaded areas (of order δx^2) become rapidly less important and in the limit as $\delta x_i \to 0$ vanish altogether. Then,

$$A = \lim_{\delta x_i \to 0} \sum_{i=1}^{N \to \infty} y_i\,\delta x_i = \int_{x_1}^{x_N} y\ dx.$$

Here the integration is regarded as the limit of a summation (i.e. view (ii)). In fact the integral sign $\int$ is an elongated 's' standing for 'summation'. This view of integration provides the route to finding areas and centres of gravity of laminae (thin plates) and three-dimensional solids.

Fig. 3 Integration consider as a limit of summation of areas

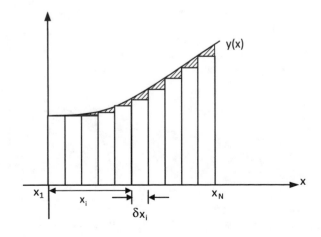

2 Areas and Centres of Area of Laminæ

Imagine a lamina of uniform thickness and material density bounded by a closed curve c, represented by upper curve $y_2(x)$ and lower curve $y_1(x)$, and of weight w per unit area (see Fig. 4).

Cut this area into narrow strips parallel to the y-axis. Let one such strip lie between x_i and $x_i + \delta x_i$.

The CG of the strip is clearly at its mid-height point, i.e. at

$$y_i = \frac{1}{2}[y_2(x_i) + y_1(x_i)].$$

The area of the strip is,

$$\delta A = [y_2(x_i) - y_1(x_i)]\delta x_i$$

and its weight,

$$w_i = w[y_2(x_i) - y_1(x_i)]\delta x_i.$$

We can write the weight of the lamina as (see Eq. 1 in Chap. 2):

$$W = \sum_{i=1}^{N} w_i = \sum_{i=1}^{N} w[y_2(x_i) - y_1(x_i)]\delta x_i$$

and the coordinates of the Centre of Gravity as (see Eq. 9 in Chap. 2):

$$\bar{x} = \frac{\sum_{i=1}^{N} w_i x_i}{\sum_{i=1}^{N} w_i} = \frac{1}{W} \sum_{i=1}^{N} w[y_2(x_i) - y_1(x_i)]x_i \delta x_i$$

Fig. 4 Integration with variable upper and lower limits

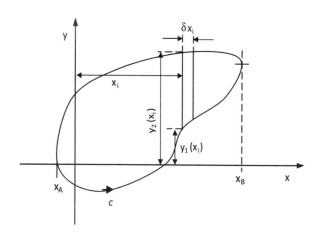

$$\overline{y} = \frac{\sum\limits_{i=1}^{N} w_i\, y_i}{\sum\limits_{i=1}^{N} w_i} = \frac{1}{W} \sum\limits_{i=1}^{N} w[y_2(x_i) - y_1(x_i)]\frac{[y_2(x_i) + y_1(x_i)]}{2}\delta x_i$$

or,

$$\overline{y} = \frac{1}{W} \sum\limits_{i=1}^{N} \frac{w}{2}[y_2(x_i)^2 - y_1(x_i)^2]\delta x_i.$$

If we now consider the limiting summations as, $\delta x_i \to 0$ the above equations become,

$$W = w \int\limits_{x_A}^{x_B}[y_2(x)] - y_1(x)]\, dx = w\, A$$

where A is the lamina area.

$$\overline{x} = \frac{w}{W} \int\limits_{x_A}^{x_B}[y_2(x) - y_1(x)]x\, dx$$

and

$$\overline{y} = \frac{w}{2W} \int\limits_{x_A}^{x_B}[y_2(x)^2 - y_1(x)^2]dx.$$

The area and coordinates of the CG can equally well be written as

$$A = \int\limits_{x_A}^{x_B}[y_2(x) - y_1(x)]\, dx \qquad\qquad (1)$$

$$\overline{x} = \frac{1}{A} \int\limits_{x_A}^{x_B}[y_2(x) - y_1(x)]x\, dx \qquad\qquad (2)$$

and

$$\overline{y} = \frac{1}{2A} \int\limits_{x_A}^{x_B}[y_2(x)^2 - y_1(x)^2]\, dx \qquad\qquad (3)$$

In this form, $\overline{x}, \overline{y}$ are properties of lamina shape and area and have nothing to do with weight or gravity per se.

The term **Centre of Gravity** or **CG** is reserved for properties of bodies possessing mass and hence weight. Where we are looking at properties of an area or volume, which have nothing to do with mass and weight directly, the term **Centroid** is used rather than Centre of Gravity.

3 A Simple Example

Consider a lamina symmetric about the x-axis defined by (see Fig. 5)

$$y_2 = a\,x^n \qquad y_1 = -y_2$$

Equation (1) gives

$$A = 2a \int_{x_A}^{x_B} x^n dx = \frac{2a}{n+1}[x^{n+1}\,]_{x_A}^{x_B}$$

or

$$A = \frac{2a}{n+1}[x_B^{n+1} - x_A^{n+1}] \tag{4}$$

Equation (2) gives

$$\bar{x} = \frac{2a}{A}\int_{x_A}^{x_B} x^n x\,dx = \frac{2a}{A}\int_{x_A}^{x_B} x^{n+1} dx$$

or

$$\bar{x} = \frac{2a}{A(n+2)}[x^{n+2}\,]_{x_A}^{x_B},$$

$$= \frac{2a}{A(n+2)}[x_B^{n+2} - x_A^{n+2}].$$

Fig. 5 A simple laminar with edges approximated by a generalised power of x

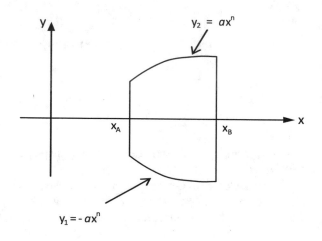

On substituting for A this yields

$$\overline{x} = \left(\frac{n+1}{n+2}\right)\left[\frac{x_B^{n+2} - x_A^{n+2}}{x_B^{n+1} - x_A^{n+1}}\right] \tag{5}$$

Clearly from symmetry $\overline{y} = 0$ in this case.

Note carefully that $y_2(x)$ and $y_1(x)$ are **not** constants and therefore **cannot** be taken outside the integration, nor may you substitute mean values of y_2 and y_1 before integrating. Particular values of n are of interest:

3.1 Rectangular Lamina ($n = 0$)

$y_2 = a$ (constant) and the lamina is a rectangle for which,
$A = 2a[x_B - x_A]$ = breadth $(2a)$ length $(x_B - x_A)$ and

$$\overline{x} = \frac{1}{2}\frac{(x_B^2 - x_A^2)}{(x_B - x_A)} = \frac{1}{2}\frac{(x_B + x_A)(x_B - x_A)}{(x_B - x_A)} = \frac{x_B + x_A}{2}.$$

[Both results pretty obvious in this case].

3.2 Triangular Lamina ($n = 1$)

$x_A = 0$ (see Fig. 6) and from Eq. (4)

$$A = \frac{2a}{2}[x_B^2 - x_A^2] = a\,x_B^2 = \frac{1}{2}b\ x_B$$

Fig. 6 Area and centre gravity calculation for a triangular lamina

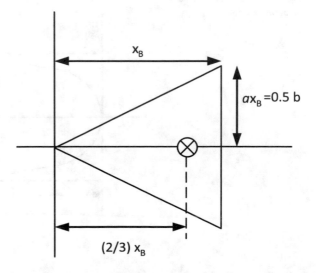

x_B

$ax_B = 0.5\ b$

$(2/3)\ x_B$

and from Eq. (5)

$$\bar{x} = \frac{2}{3}\frac{x_B^3 - x_A^3}{x_B^2 - x_A^2} = \frac{2}{3}x_B.$$

Hence for a triangle the centroid is at 2/3 *height* from the *apex* or 1/3 *height* from *base*.

3.3   Parabolic Lamina I, $(n = 1/2)$

$x_A = 0, \; x_B = L$

In this case the lamina is parabolic since $y = ax^{\frac{1}{2}}$ is equivalent to $y^2 = a^2 x$ which is the equation of a parabola as sketched in Fig. 7.

From Eq. (4), in this case,

$$A = \frac{2a}{\frac{3}{2}}[x_B^{3/2} - x_A^{3/2}] = \frac{4a}{3}x_B \; x_B^{1/2} = \frac{2}{3}L \;\; B$$

Thus the area of the parabolic lamina is 2/3 the area of the surrounding rectangle. From Eq. (5)

$$\bar{x} = \frac{\frac{3}{2}}{\frac{5}{2}}\left[\frac{x_B^{5/2} - x_A^{5/2}}{x_B^{3/2} - x_A^{3/2}}\right] = \frac{3}{5}x_B = \frac{3}{5}L.$$

Thus the centroid is $\frac{3}{5}L$ from $x = 0$.

Fig. 7 Area of a blunt-nosed parabolic laminar

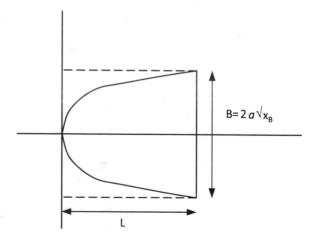

Fig. 8 Area of pointed nosed parabolic laminar

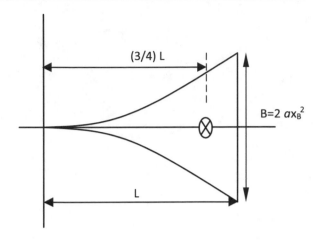

3.4   Parabolic Lamina II, $(n = 2)$

$x_A = 0, x_B = L$.

In this case $y_2 = ax^2$, which is also parabolic, but representing a different shape (see Fig. 8). From Eq. (4)

$$A = \frac{2a}{3}x_B^3 = \frac{1}{3}L \ B.$$

From Eq. (5)

$$\bar{x} = \frac{3}{4}\frac{x_B^4}{x_B^3} = \frac{3}{4}L.$$

There are clearly many other possible cases. Also different forms of equation for $y_1(x)$ and $y_2(x)$ can be considered. However, the treatment is similar in all cases. It simply involves properly integrating equations (1), (2) and (3).

4   Summary

1. The basic principle of calculating geometric properties of displacement volume through the use of sectional area is introduced.
2. The concept of integration is summarised.
3. Area and centroid coordinates for a general lamina are derived, using the area and relevant (first) moments of an infinitesimally small strip and subsequently integrating along the lamina.
4. The centroid coordinates are obtained by dividing the relevant (first moment) of area by the area.
5. Examples are given for simple geometric shapes: rectangle, triangle and parabolas.

Further Comments on Displacement Volume and Centre of Buoyancy

5

1 Calculation of Displacement and Centre of Buoyancy

We will adopt the axis system with origin at amidships centre plane and base line, as shown in Fig. 1, and assume, for the moment, that our floating body, or ship, is symmetric about the centreline (xz) plane. This will imply that the centre of buoyancy is on centreline (i.e. $y = 0$).

Volume and moment of volume properties can be found in two ways:

1. Cut the vessel into transverse slices along the vessel length (see Fig. 2)
2. Cut the vessel into horizontal slices or waterline slices throughout the vessel draught (see Fig. 2).

In case (1) let the transverse sectional area be $a(x)$ to the required draught. Then the displacement volume is

$$\nabla = \int_{x_A}^{x_F} a(x)dx. \tag{1}$$

The moment of volume about $x = 0$ [amidships ($\otimes$)] is

$$M_x = \int_{x_A}^{x_F} a(x)\, x\, dx = \nabla\, \bar{x} \tag{2}$$

where $\bar{x}$ is the **longitudinal** position of the **centre of buoyancy** (LCB).

These equations follow from the element properties,

$$\delta\nabla = a(x)\, \delta x \tag{3}$$

$$\delta M_x = \delta\nabla\, x = a(x)\, x\, \delta x. \tag{4}$$

© Springer International Publishing AG, part of Springer Nature 2018
P. A. Wilson, *Basic Naval Architecture*,
https://doi.org/10.1007/978-3-319-72805-6_5

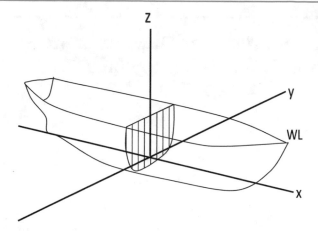

Fig. 1 Volume estimation by approximation of ship by longitudinal sections

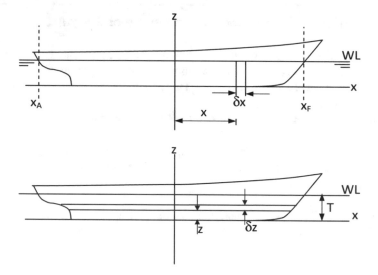

Fig. 2 Volume approximation of ship by using waterplane areas

In case (2) let the waterplane area at z be $A(z)$. Then the displacement volume ∇ is

$$\nabla = \int_0^T A(z)\, dz \tag{5}$$

and the moment of volume about $z = 0$ (baseline) is

$$M_z = \int_0^T A(z)\, z\, dz = \nabla\, \bar{z} \tag{6}$$

where $\bar{z}$ defines the **vertical** position of the **centre of buoyancy** (VCB).

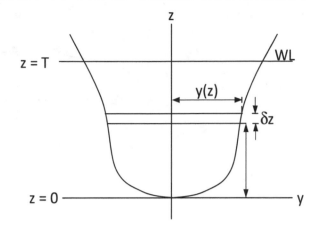

Fig. 3 Section at distance x from amidships

2 Calculation of Sectional Area

To complete the calculations for case (1) it is necessary to calculate sectional areas $a(x)$ from the waterline half breadths or offsets $y(z, x)$ (see Fig. 3).

Thus

$$a(x) = 2 \int_0^T y(z, x)dz. \tag{7}$$

Note y is written as $y(z, x)$ because it is a function of **both** fore **and** aft section position and vertical draught. The factor 2 is because the full width of ship is $2y(z, x)$. Missing out this 2 *factor* is the commonest calculation error in naval architecture. Plots showing the variation of sectional area with draught for all stations along the vessel are a common feature of *hydrostatics*; such plots are, usually called **Bonjean** curves.

3 Calculation of Waterplane Area and Centroid

Likewise to complete the calculations for case (2) waterplane properties are needed (see Fig. 4).

In this case,

$$A(z) = 2 \int_{x_A}^{x_F} y(z, x)dx \tag{8}$$

and also,

$$A(z)\, x'(z) = 2 \int_{x_A}^{x_F} y(z, x)\, x\, dx \tag{9}$$

where $x'(z)$ is the position of the **centre of area** (centroid) of the given waterplane.

Fig. 4 Waterplane area
estimation from offset data

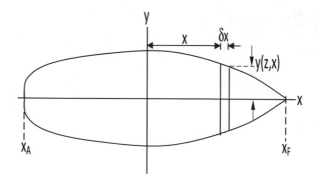

4 Introduction to Changes of Draught (Parallel Sinkage) and Trim

If there is a small increase of draught δT, with the new WL parallel to the old WL, there will be additional volume and buoyancy. The additional buoyancy force will act through the centre of the added layer, that is, through the *centre of area* of the water plane.

Any additional mass placed on board will need to be put over the centroid of the waterplane in order to produce a **parallel sinkage** (see Fig. 5). If the mass is placed ahead of this centre the vessel will trim bow down; if behind (or abaft) this centroid it will **trim** stern down.

Trim is defined as,

$$t = T_F - T_A \tag{10}$$

where $T_F =$ Draught at forward perpendicular and $T_A =$ Draught at aft perpendicular.

Fig. 5 Definition of parallel
sinkage

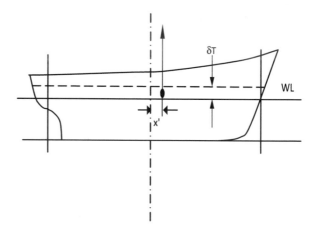

Mean draught is,

$$T_M = \frac{T_F + T_A}{2} \tag{11}$$

which is draught at amidships (between perpendiculars).

Because of its significance in respect to changes of trim the centre of area of the waterplane is called the centre of flotation (LCF).

LCF = Longitudinal position of the centre of flotation.

On changing draught by parallel sinkage the change of displacement is

$$\delta\Delta = \rho \, A \, \delta T \text{(mass)} \tag{12}$$

or

$$\delta\Delta = \rho \, g \, A \, \delta T \text{(weight)}. \tag{13}$$

The derivative values,

$$\frac{d\Delta}{dT}, \tag{14}$$

indicate the rate of change of displacement with draught T, or change of displacement per unit draught change.

In mass units this is commonly expressed as:

tonnes per metre (TPM) or tonnes per cm (TPC).

5 Movement of LCB Due to a Small Change of Trim

If there is a change of trim due to small changes of draught the angular rotation of the vessel is given by (see Fig. 6)

$$tan\theta = \frac{\delta \, T_F - \delta \, T_A}{L_{BP}} \approx \theta \, (radians) \qquad \theta \text{ is small} \tag{15}$$

Any change of draught can be thought of as taking place in two stages, as shown in Fig. 6:

Fig. 6 Trim change on waterline

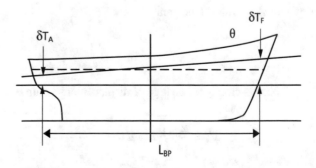

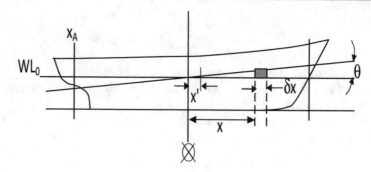

Fig. 7 Calculation of trimmed *LCF*

1. A parallel sinkage without change of trim, followed by
2. A change of trim at constant displacement.

For case (2) the original WL_0 and the new trimmed WL_1 will intersect at, say, $x = x'$ (see Fig. 7). The change of draught at x is (for small θ)

$$(x - x') \, \theta. \tag{16}$$

Thus, the shaded element of volume is,

$$\delta V = 2y \, (x - x') \, \theta \, \delta x. \tag{17}$$

The element will have a moment of volume about **amidships** of

$$\delta m_x = x \, \delta V = 2y \, x \, (x - x') \, \theta \, \delta x. \tag{18}$$

By integrating over the whole volume between the old and new waterlines the change of displacement volume is obtained as

$$\delta V = 2 \int_{x_A}^{x_F} y \, (x - x') dx \, \theta = \left\{ 2 \int_{x_A}^{x_F} x \, y \, dx - 2x' \int_{x_A}^{x_F} y \, dx \right\} \theta = \left\{ 2 \int_{x_A}^{x_F} x \, y \, dx - x' \, A \right\} \theta \tag{19}$$

where A = waterplane area

$$2 \int_{x_A}^{x_F} y \, dx. \tag{20}$$

There is clearly no change of displacement volume if x' is chosen so that,

$$x' = \frac{2}{A} \int_{x_A}^{x_F} x \, y \, dx = \frac{First \ Moment \ of \ WPArea}{WPArea}. \tag{21}$$

This means that x' is the position of the centre of area of the waterplane. Thus,

If a vessel trims at constant displacement it does so about an axis through the centre of flotation (i.e. about the LCF).

Obviously the longitudinal centre of buoyancy (LCB) will move forward as the vessel trims bow down. This can be calculated from the change of moment of volume due to change of trim:

$$\delta M_x = \nabla \, \delta \bar{x} = \int_{x_A}^{x_F} \delta m_x = 2 \int_{x_A}^{x_F} y \, x \, (x - x') \, dx \, \theta \quad . \tag{22}$$

Thus the forward movement of LCB is

$$\delta \bar{x} = \frac{1}{\nabla} \left\{ 2 \int_{x_A}^{x_F} x^2 \, y \, dx \; - \; 2 \, x' \int_{x_A}^{x_F} x \, y \, dx \right\} \theta = \frac{1}{\nabla} \left\{ 2 \int_{x_A}^{x_F} x^2 \, y \, dx \; - \; A \, (x')^2 \right\} \theta \tag{23}$$

since,

$$2 \int_{x_A}^{x_F} x \, y \, dx = A \, x' \quad . \tag{24}$$

It can be seen that there is now a need to calculate integrals of the form,

$$J_{0L} = 2 \int_{x_A}^{x_F} x^2 \, y \, dx \quad . \tag{25}$$

6 Longitudinal Second Moments of Area and Parallel Axis Theorem

For the waterplane element shown in Fig. 4,

the area is $\qquad\qquad \delta A = 2y\delta x$ and

the First Moment of Area is $\qquad x \times \delta A = 2xy\delta x$.

This scheme can be extended to higher order moments by defining

Second Moment of Area as $\qquad x^2 \times \delta A = 2\,x^2 y \, \delta x$.

The reason for calling the ordinary *moment of area* the First Moment is now clear. Thus,

$$J_{0L} = 2 \int_{x_A}^{x_F} x^2 \, y \, dx \tag{26}$$

is **Longitudinal Second Moment of Area** about $x = 0$; hence subscripts $0, L$.

A Second Moment of Area could also be defined about an axis through LCF (see Fig. 7) as,

$$J_L = 2 \int_{x_A}^{x_F} (x - x')^2 \ y \, dx \tag{27}$$

or

$$J_L = 2 \int_{x_A}^{x_F} [x^2 - 2x' \ x + (x')^2] \ y \, dx \tag{28}$$

or

$$J_L = 2 \int_{x_A}^{x_F} x^2 \ y \, dx \ - \ 2 \ x' \ 2 \int_{x_A}^{x_F} x \ y \, dx + (x')^2 \ 2 \int_{x_A}^{x_F} y \, dx \tag{29}$$

$$J_L = J_{0L} - 2x' \ A \ x' + (x')^2 \ A. \tag{30}$$

That is to say,

$$J_L = J_{0L} - A(x')^2 \tag{31}$$

or, equally well,

$$J_{0L} = J_L + A(x')^2. \tag{32}$$

This relation is called **THE PARALLEL AXIS THEOREM**. This relation makes it clear, since, $A \ (x')^2$ is always positive, that the second moment of area about an axis through the centre of area (or centroid) J_L is the minimum possible second moment. This equation also provides a simple method of correcting the second moment of area for an axis a distance x' from the centre of area.

7 Formulæ for LCB Shift and Longitudinal Metacentre

The LCB shift due to change trim, Eq. 23, can be written as,

$$\delta \bar{x} = \frac{1}{\nabla} \left\{ 2 \int_{x_A}^{x_F} x^2 \ y \, dx - A(x')^2 \right\} \ \theta = \frac{1}{\nabla} \left\{ J_{0L} - A(x')^2 \right\} \ \theta. \tag{33}$$

That is to say,

$$\delta \bar{x} = \frac{J_L}{\nabla} \ \theta. \tag{34}$$

Fig. 8 Definition of
longitudinal metacentre M_L

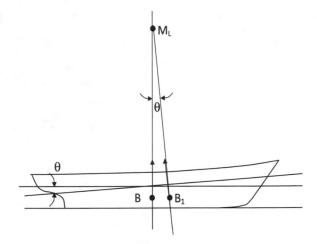

At the old waterline the buoyancy Δ acts through B, perpendicular to the old
waterline. After trimming Δ acts through B_1, perpendicular to the new waterplane
(see Fig. 8). The distance,

$$BB_1 = \delta \bar{x} = \frac{J_L}{\nabla} \, \theta. \tag{35}$$

The two lines of action of Δ intersect at M_L where the distance BM_L is (for small
θ) such that,

$$\text{BM}_\text{L} \, \theta = \text{BB}_1 = \frac{J_L}{\nabla} \, \theta \tag{36}$$

Thus,

$$BM_L = \frac{J_L}{\nabla}. \tag{37}$$

M_L is identified as the **longitudinal metacentre** for the vessel floating at the
given waterline, and the formula (which should be committed to memory!) gives the
height of this metacentre vertically above the centre of buoyancy.

8   Movement of Centre of Buoyancy Due to Small Heel Angle

Consider the section, representing a slice of length δx, shown in Fig. 9 where

$$W_o L_o = \text{Upright Waterplane} \tag{38}$$

$$W_1 L_1 = \text{Heeled Waterplane}. \tag{39}$$

Fig. 9 Effects of heel
assuming wall sided ship
section

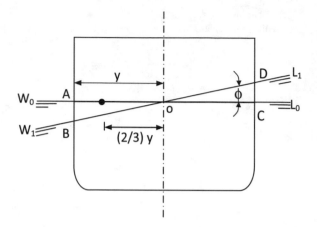

As the vessel heels wedge OAB emerges (comes out of water) and wedge OCD immerses (goes in). Thus volume of either wedge is,

$$\delta \nabla = \frac{1}{2} \, y \, (y \; \varphi) \, \delta x \tag{40}$$

since for small φ (rads) $AB = DC = y\varphi$.

Centroid of each wedge is at $\frac{2}{3} y$ from O (centroid of a triangle).

Thus there is a transfer of moment of volume from the emerging side to the immersing side given by:

$$\delta m_y = 2 \, \frac{2}{3} \, y \, \frac{1}{2} y^2 \, \delta x \, \varphi = \frac{2}{3} \, y^3 \, \delta x \; \varphi. \tag{41}$$

Integrating along the ship length there is a total change of moment of volume given by,

$$\delta M_y = \frac{2}{3} \int_{x_A}^{x_F} y^3 \, dx \; \varphi = J_T \, \varphi. \tag{42}$$

As a result of this transfer of moment of volume the centre of buoyancy moves transversely from B (on Centreline for symmetric vessel) to B_1 such that (see Fig. 10)

$$BB_1 = \frac{J_T}{\nabla} \; \varphi. \tag{43}$$

This is analogous to the fore and aft shift of LCB due to change of trim. The new and old lines of action of buoyancy intersect at M_T where, again analogous to the trim case, the height of the point M_T above the upright centre of buoyancy is given by, for small, as $BM_T = BB_1$ and, thus,

$$BM_T = \frac{J_T}{\nabla}. \tag{44}$$

Fig. 10 Transverse
metacentre definition

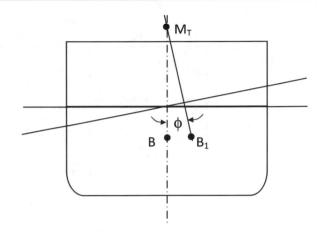

M_T is the **transverse metacentre** for the vessel at the given WL. J_T is identified in the next section of these notes as the transverse second moment of area about the fore and aft centreline, it is evaluated as,

$$J_T = \frac{2}{3} \int_{x_A}^{x_F} y^3 \, dx. \tag{45}$$

9   Second Moments of Area of Simple Laminæ

9.1   Rectangular Lamina

For a simple rectangle the longitudinal second moment of area is found as follows (see Fig. 11):
using

$$\delta J_L = x^2 B \, \delta x \tag{46}$$

then

$$J_L = B \int_{\frac{-L}{2}}^{\frac{L}{2}} x^2 \, dx = B \left[\frac{x^3}{3} \right]_{-L/2}^{L/2}. \tag{47}$$

Thus,

$$J_L = \frac{BL^3}{12}. \tag{48}$$

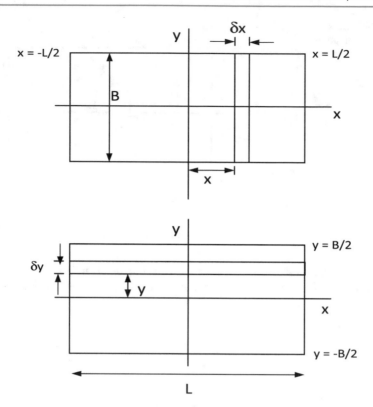

Fig. 11 Calculation of J_T for a rectangular laminar

The transverse Second Moment can be written in one of two ways, either (see Fig. 11): using,

$$\delta J_T = y^2 \, L \, \delta y, \tag{49}$$

then,

$$J_T = L \int_{\frac{-B}{2}}^{\frac{B}{2}} y^2 \, dy = L \left[\frac{y^3}{3} \right]_{-B/2}^{B/2}, \tag{50}$$

or, alternatively (see Fig. 11), taking the second moment of the slice x about x-axis,

$$\delta J_T = \left(\frac{B^3}{12} \right) \delta x \tag{51}$$

then

$$J_T = \frac{1}{12} \int_{-L/2}^{L/2} B^3 \, dx = \left[\frac{B^3}{12} x \right]_{-L/2}^{L/2}, \tag{52}$$

Fig. 12 Typical ship
waterplane used for
calculation of J_T

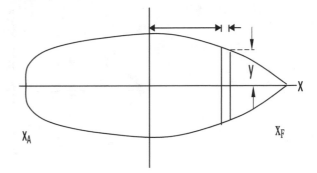

either way

$$J_T = \frac{L \; B^3}{12}.$$ (53)

9.2   Typical Ship Waterplane

This simple result justifies the general formula for a waterplane (see Fig. 12):
For the element of length x and height $2y$ (approximately a rectangle),

$$\delta \; J_T = \frac{1}{12} \; (2y)^3 \; \delta x = \frac{2}{3} y^3 \; \delta x,$$ (54)

and integrating for the whole waterplane.

$$J_T = \frac{2}{3} \int_{x_A}^{x_F} y^3 \; dx,$$ (55)

as required for the estimation of,

$$B M_T.$$ (56)

9.3   Mathematically Defined Waterplane

Next consider a simple curve of the form,

$$y = \frac{B}{2} \left(1 - \left(\frac{2x}{L}\right)^n \right)$$ (57)

extending from $x = 0$ to $x = L/2$. Consider a complete waterplane formed by
reflecting this curve about the y-axis (see Fig. 13). Note that $y = 0$ at $x = L/2$ and

Fig. 13 Mathematically
defined waterplane

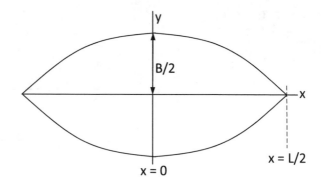

$y = B/2$ at $x = 0$. Due to assumed symmetry the area A is

$$A = 2 \int_{\frac{-L}{2}}^{\frac{L}{2}} y \, dx = 4 \int_0^{\frac{L}{2}} y \, dx = 4 \frac{B}{2} \int_0^{\frac{L}{2}} \left(1 - \left(\frac{2x}{L}\right)^n\right) dx$$

$$= 2B \left[x - \left(\frac{2}{L}\right)^n \frac{x^{n+1}}{n+1}\right]_0^{\frac{L}{2}} = LB \left\{1 - \frac{1}{n+1}\right\}; \tag{58}$$

the longitudinal second moment of area is:

$$J_L = 4 \int_0^{\frac{L}{2}} x^2 \, y \, dx = 2B \int_0^{\frac{L}{2}} \left(x^2 - \left(\frac{2}{L}\right)^n x^{n+2}\right) dx \tag{59}$$

$$J_L = 2B \left[\frac{x^3}{3} - \left(\frac{2}{L}\right)^n \frac{x^{n+3}}{n+3}\right]_0^{\frac{L}{2}} \tag{60}$$

giving,

$$J_L = \frac{L^3 B}{4} \left(\frac{1}{3} - \frac{1}{n+3}\right) . \tag{61}$$

Using the algebraic identity,

$$(1 + a)^3 = 1 + 3a + 3a^2 + a^3, \tag{62}$$

the transverse second moment of area can be evaluated as:

$$J_T = 2 \frac{2}{3} \int_0^{L/2} \left(\frac{B}{2}\right)^3 \left(1 - \left(\frac{2x}{L}\right)^n\right)^3 dx = \frac{B^3}{6} \int_0^{L/2} \left(1 - 3\left(\frac{2x}{L}\right)^n + 3\left(\frac{2x}{L}\right)^{2n} - \left(\frac{2x}{L}\right)^{3n}\right) dx \tag{63}$$

$$J_T = \frac{B^3}{6} \left[x - 3 \left(\frac{2}{L}\right)^n \frac{x^{n+1}}{n+1} + 3 \left(\frac{2}{L}\right)^{2n} \frac{x^{2n+1}}{2n+1} - \left(\frac{2}{L}\right)^{3n} \frac{x^{3n+1}}{3n+1} \right]_0^{L/2}$$

(64)

or

$$J_T = \frac{L\,B^3}{12} \left[1 - \frac{3}{(n+1)} + \frac{3}{(2n+1)} - \frac{1}{(3n+1)} \right].$$

(65)

By choosing the index n appropriately waterlines can be generated having a wide range of waterplane area coefficients C_W. These can give very realistic estimates of J_T and J_L for normal ship waterplanes. The processes used above can clearly be extended for other classes of curve defined by different functions $y(x)$. For a parabolic waterline $n = 2$, and hence,

$$A = \frac{2}{3} L\,B, \quad J_L = \frac{L^3\,B}{30}, \quad J_T = \frac{4\,L\,B^3}{105} \quad !$$

(66)

a result which you can verify.

9.4   Circular Lamina

Now consider the Second Moment of Area of a circle about a diameter (See Fig. 14):

$$J_L = 2 \int_{-R}^{R} x^2\,y\,dx.$$

(67)

Using polar coordinates

$$x = R\,\cos\theta \quad ; \quad dx = -R\,\sin\theta\,d\theta$$

(68)

Fig. 14 Second moment of area, J_T and J_L for a circular laminar

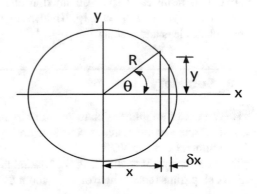

and

$$y = R \, \sin\theta. \tag{69}$$

Thus

$$J_L = 2 \int_\pi^0 R^2 \, \cos^2\theta R \, \sin\theta(-R \, \sin\theta \, d\theta) = 2R^4 \int_0^\pi \cos^2\theta \, \sin^2\theta \, d\theta \tag{70}$$

Since,

$$\sin 2\theta = 2\cos\theta \sin\theta \tag{71}$$

$$\sin^2 2\theta = 4\cos^2\theta \sin^2\theta \tag{72}$$

$$J_L = \frac{R^4}{2} \int_0^\pi \sin^2 2\theta d\theta \tag{73}$$

also

$$\cos 4\theta = 1 - 2\sin^2 2\theta \tag{74}$$

as

$$J_L = \frac{R^4}{2} \int_0^\pi (1 - \cos 4\theta) d\theta \tag{75}$$

This can be integrated to give:

$$J_L = \frac{R^4}{4} \left[\theta - \frac{1}{4}\sin 4\theta \right]_{\theta=0}^{\pi}. \tag{76}$$

Hence,

$$J_L = \frac{\pi R^4}{4} = \frac{\pi D^4}{64}. \tag{77}$$

as $D = 2R$. Naturally the same result applies for the transverse second moment of area. This result can also be obtained through obtaining the polar second moment of Area $J_{polar} = 2J_L = 2J_T$. This result is useful for offshore structures with cylindrical legs and bracing.

10 Summary

1. The use of an infinitesimally small strip and subsequent integration is extended to define sectional area, waterplane area and centroid (LCF) and displacement volume, LCB and VCB.
2. The concepts of small changes of draught and trim and heel are introduced.
3. A ship trims (small angle of trim) about LCF without changing its displacement.

4. Movement of centre of buoyancy for small angles of trim and heel is derived using relevant (first) moments.
5. Longitudinal and transverse second moments of area are defined and illustrated for various shapes.
6. The parallel axis theorem is defined, stating that the smallest second moment of area is about an axis through the centroid.
7. The positions of transverse and longitudinal metacentre, with reference to VCB, are defined using the relevant second moments of area as:

$$BM_{L,T} = \frac{J_{L,T}}{\nabla}.$$

(78)

Numerical Integration Formulæ

In ships and other floating structures calculation of section areas, waterplane areas, volumes and various properties such as first and second moments is commonplace. This leads to the evaluation of integrals of the form seen in the previous Chap. 5, for example,

$$A = 2 \int_{x_A}^{x_F} y(x)\, dx, \tag{1}$$

$$M_x = 2 \int_{x_A}^{x_F} x\, y(x)\, dx, \tag{2}$$

$$J_{xx} = 2 \int_{x_A}^{x_F} x^2 y(x)\, dx, \tag{3}$$

$$a(x) = 2 \int_{0}^{T} y(z, x)\, dz, \tag{4}$$

$$\nabla = \int_{x_A}^{x_F} a(x)\, dx, \tag{5}$$

etc.

Where x is measured along the hull with x_F and x_A indicating the fore and aft limits, $y(x)$ is the half-breadths or offsets of a port/starboard symmetric waterplane, A the waterplane area and M_x and J_{xx} its longitudinal first and second moments of

© Springer International Publishing AG, part of Springer Nature 2018
P. A. Wilson, *Basic Naval Architecture*,
https://doi.org/10.1007/978-3-319-72805-6_6

Fig. 1 Function evaluated at
points spaced at even
intervals

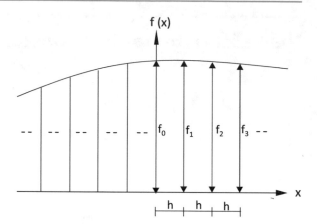

area about the chosen origin. $y(z, x)$ denotes the half-breadths between the keel and
the load waterline at a particular station along the ship. Similarly ∇ is the underwater
volume and $a(x)$ the underwater sectional area.

The integrands $y(x)$, $xy(x)$, $x^2y(x)$, $a(x)$, etc., are not, in general, analytical functions. Their values are, however, available at various points along the hull, i.e. stations, or along the vertical axis, i.e. waterlines. Therefore, to evaluate these integrals
numerical integration of a function $f(x)$ or $f(z)$ is required where values of the
function are available at discrete points f_0, f_1, f_2, etc., as shown in Fig. 1.

With any numerical integration formula it is important to know the accuracy of
the result. The sources of inaccuracy can be attributed to rounding errors associated
with the measurements of the function to be integrated (typically measurements
by eye with a ruler are accurate to 0.5 mm and thus are rounded to an appropriate
number of decimals) and truncation errors relating to the analytical approximation
used to represent the function to be integrated. Typically a function is represented in
polynomial form,

$$f(x) = a_0 + a_1 x + a_2 x^2 + a_3 x^3 + \ldots\ldots \tag{6}$$

suitably truncated or cut off after the first few terms.

1 Trapezoidal Rule

This is the simplest form of integration involving two ordinates and assuming the
variation of the function between these two ordinates is linear, as shown by the dotted
line in Fig. 2. Selecting a 2nd order polynomial (in order to find the truncation error)
and placing the origin at $x = 0$

$$f(x) = f_0 + a x + b x^2 \tag{7}$$

Fig. 2 Trapezoidal rule

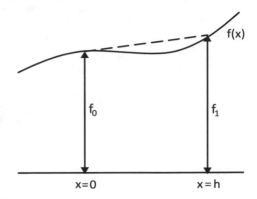

such that [from $f(x = h) = f_1$]

$$a\,h = f_1 - f_0 - b\,h^2 \tag{8}$$

where h is the interval between the ordinates. The integral is,

$$I = \int_0^h f(x)\,dx = f_0\,h + \frac{1}{2}\,a\,h^2 + \frac{1}{3}\,b\,h^3 = \frac{f_0 + f_1}{2}\,h - \frac{1}{6}\,b\,h^3 = I_{TI} + \varepsilon_{tr} \tag{9}$$

where I_{TI} represents the value of the integral assuming the area under $f(x)$, between $x = 0$ and $x = h$, can be represented by the area of a trapezoid and t_r is the truncation error. Thus the trapezoidal rule for a function defined at $N+1$ points (i.e. N intervals of length h) is

$$I_{TI} = h\left(\frac{f_0}{2} + f_1 + f_2 + f_3 + \dots\dots + f_{N-1} + \frac{f_N}{2}\right) \tag{10}$$

The truncation error is proportional to the second derivative of the function $f(x)$, as $b = f''/2$, and depends on h^3, i.e. the total number of intervals used in the numerical integration.

2 Simpson's First Rule

The function between $-h \leq x \leq h$ is represented by a 2nd order polynomial. The function is described by three ordinates, namely f_{-1}, f_0 and f_1 (see Fig. 3). Using a 4th order polynomial (in order to find the truncation error),

$$f(x) = f_0 + a\,x + b\,x^2 + c\,x^3 + d\,x^4 \tag{11}$$

Fig. 3 Simpson's first rule

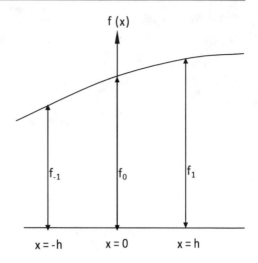

the integral can be expressed as,

$$\int_{-h}^{h} f(x)\, dx = \left[f_0\, x + \frac{1}{2}\, a\, x^2 + \frac{1}{3}\, b\, x^3 + \frac{1}{4}\, c\, x^4 + \frac{1}{5}\, d\, x^5 \right]_{-h}^{h}$$

$$= 2\, f_0\, h + \frac{2}{3}\, b\, h^3 + \frac{2}{5}\, d\, h^5. \tag{12}$$

From Eq. 11 and referring to Fig. 3 firstly with $x = -h$,

$$f_{-1} = f(x = -h) = f_0 - a\, h + b\, h^2 - c\, h^3 + d\, h^4 \tag{13}$$

and then with $x = -h$,

$$f_1 = f(x = h) = f_0 + a\, h + b\, h^2 + c\, h^3 + d\, h^4 \tag{14}$$

thus by adding Eqs. 13–14 leading to,

$$f_1 + f_{-1} = 2\, f_0 + 2\, b\, h^2 + 2\, d\, h^4 \tag{15}$$

so,

$$\frac{2}{3}\, b\, h^2 = \frac{1}{3}\, (f_1 + f_{-1}) - \frac{2}{3}\, f_0 - \frac{2}{3}\, d\, h^4. \tag{16}$$

Substituting Eq. 16 into Eq. 12 the integral becomes,

$$I = \frac{4}{3}\, f_0\, h + \frac{1}{3}\, (f_1 + f_{-1})\, h - \frac{4}{15}\, d\, h^5 = I_{S1} + \varepsilon_{tr}. \tag{17}$$

Thus, Simpson's **FIRST** rule for the area under the curve presented by three ordinates spaced at two equal intervals is,

$$I_{S1} = \frac{h}{3} \left(f_{-1} + 4 \, f_0 + f_1 \right) , \tag{18}$$

with the truncation error being proportional to the 4th derivative of function $f(x)$, i.e. $d = f^{IV}/24$, and depending on h^5. The area under a curve represented by N number of even equal intervals of length h [i.e. odd number of ordinates for describing $f(x)$], starting from f_0 and terminating at f_N, as this is more suitable for ship-related calculations, is,

$$\begin{aligned} I_{S1} = \frac{h}{3} (f_0 &+ 4 \, f_1 + 2 f_2 + 4 f_3 + 2 f_4 + 4 f_5 + \cdots + 4 f_{N-3} \\ &+ 2 f_{N-2} + 4 f_{N-1} + f_N) \end{aligned} \tag{19}$$

Here Eq. 19 was obtained by applying Eq. 18 to successive pairs of intervals and summing up the obtained areas.

The constant coefficients $1, 4, 2, 4, 2, 4, 2, 4, \ldots, 4, 2, 4, 1$ are usually referred to as the **Simpson Multipliers**.

2.1 Example

The following is an example application of Simpson's first rule in obtaining areas, centroids and second moments of the waterplane area for a hull with LBP $= 25$ m. The half-breadths at the waterline $y(x)$ are provided as input at 11 stations along the ship, 0 being the Aft Perpendicular (AP) and 10 the Fore Perpendicular (FP). Note, however, that data is also provided at half-stations between stations 0 and 1, 1 and 2, 8 and 9, 9 and 10, respectively. This is common practice and arises from the necessity to provide additional data in the regions of the ship with rapid changes in curvature. The half-breadths and corresponding Simpson Multipliers are shown in columns (2) and (3) of Table 1. There are differences in the Simpson Multipliers, by comparison to Eq. 19, due to the use of half-stations. Obtain the Simpson Multipliers, shown in column (3) of Table 1, by applying Eq. 19 to pairs of intervals with length $0.5h$ at the appropriate regions and expressing the areas as a function of h, rather than $0.5h$. These, then, are added to the area corresponding to the full stations to obtain the Simpson Multipliers shown in Table 1.

The interval to be used when applying Simpson's first rule is $h = 2.5$ m. Also note that the arm for the moments (first and second) given in column (5) in Table 1 is with reference to amidships (station 5), i.e. the origin of the axes system is at amidships with x positive forward. The final results do not change if a different reference, such as AP or FP, is selected.

Table 1 Integration using Simpson's rule in tabular form

(1)	(2)	(3)	(4) = (3)(2)	(5)	(6) = (4)(5)	(7) = (6)(5)	(8) = (2)²	(9) = (8)(3)
Station	$y(x)$	SM	$A_1(x)$	$l(x)$	$M_1(x)$	$J_1(x)$	$y^3(x)$	$J_2(x)$
	(m)		(m)	(m)	(m²)	(m³)	(m³)	(m³)
0	0.0	0.5	0.0	−12.50	0.0	0.0	0.0	0.0
0.5	0.406	2	0.812	−11.25	−9.315	102.769	0.067	0.134
1	0.792	1	0.792	−10.00	−7.920	79.20	0.497	0.497
1.5	1.142	2	2.284	−8.75	−19.985	174.869 m	1.489	2.978
2	1.442	1.5	2.163	−7.50	−16.623	121.673	2.998	4.497
3	1.840	4	7.360	−5.00	−36.800	184.000	6.230	24.920
4	1.982	2	3.964	−2.50	−9.910	24.775	7.786	15.572
5	1.946	4	7.784	0.0	0.0	0.0	7.369	29.476
6	1.758	2	3.516	2.50	8.790	21.975	5.433	10.866
7	1.414	4	5.656	5.00	28.280	141.400	2.827	11.308
8	0.900	1.5	1.350	7.50	10.125	75.938	0.729	1.094
8.5	0.618	2	1.236	8.75	10.815	94.631	0.236	0.472
9	0.354	1	0.354	10.00	3.540	35.400	0.044	0.044
9.5	0.139	2	0.278	11.25	3.128	35.190	0.003	0.006
10	0.0	0.5	0.0	12.50	0.0	0.0	0.0	0.0
Σ			37.549		−35.295	1091.82		101.864

Thus, the waterplane area, using the summation of all elements in column (4) in Table 1, is,

$$A = 2 \, \frac{2.5}{3} \, 37.459 = 62.582 \text{ m}^2 \tag{20}$$

and the longitudinal first moment of the waterplane area about the transverse axis through amidships, using the algebraic summation of all elements in column (6) of Table 1, is,

$$M_x = 2 \, \frac{2.5}{3} \, (-\,35.295) = -58.825 \text{ m}^3 \tag{21}$$

and the LCF is at,

$$\frac{M_x}{A} = (-)0.94 \text{ m} \quad \text{aft of amidships.} \tag{22}$$

The longitudinal second moment of the waterplane area about amidships, using the summation of all elements in column (7) of Table 1, is,

$$J_{0L} = 2 \, \frac{2.5}{3} \, 1091.82 = 1819.70 \text{ m}^4. \tag{23}$$

The longitudinal second moment of the waterplane area about LCF, using the parallel axis theorem, is,

$$J_L = 1819.70 - 62.582 \, (-0.94)^2 = 1764.40 \text{ m}^4. \tag{24}$$

The transverse second moment of the waterplane area, using the summation of all elements in column (9) in Table 1, is,

$$J_T = \frac{2}{3} \frac{2.5}{3} \, 101.864 = 56.591 \text{ m}^4. \tag{25}$$

3 Simpson's Second Rule

The function, in this case, is described by 4 ordinates, namely f_1, f_2, f_3 and f_4 between $-3h/2 \le x \le 3h/2$, as shown in Fig. 4. Selecting the 4th order polynomial shown in Eq. 11, where f_0 is not in this case a measured ordinate of the function, the integral is obtained as,

$$\int_{-3h/2}^{3h/2} f(x) \, dx = 3 \, f_0 \, h + \frac{9}{4} \, b \, h^3 + \frac{243}{80} \, d \, h^5 \tag{26}$$

Using Eq. 11, at $x = -3h/2, -h/2, h/2$ and $3h/2$, it can be shown that,

$$f_1 + f_4 = 2 f_0 + \frac{9}{2} \, b \, h^2 + \frac{81}{8} \, d \, h^4 \tag{27}$$

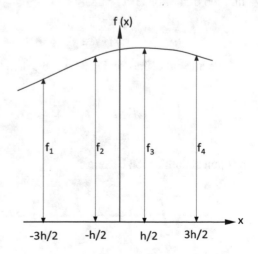

Fig. 4 Simpson's second rule for even-spaced intervals

and,

$$f_2 + f_3 = 2f_0 + \frac{1}{2} b h^2 + \frac{1}{8} d h^4. \tag{28}$$

Evaluating f_0 and bh^2 from these equations and substituting in Eq. 26 the integral becomes,

$$I = \frac{18}{16} (f_2 + f_3) h + \frac{6}{16} (f_1 + f_4) h - \frac{9}{10} d h^5 = I_{S2} - \varepsilon_{tr}. \tag{29}$$

Thus Simpson's **SECOND** rule, for the area under a curve defined by 4 ordinates at **THREE** equal intervals is,

$$I_{S2} = \frac{3}{8} h (f_1 + 3f_2 + 3f_3 + f_4) \tag{30}$$

with the truncation error being proportional to $d = f^{IV}/24$ and depending on h^5. As can be seen the truncation error associated with the second rule is more than three times of the error of the first rule, for the same value of h. This rule is suitable when the number of intervals is divisible by 3.

4 5, +8, −1 Rule

This rule is, in general, used for integrating over one segment of a curve, given by two ordinates, whilst making use of a third ordinate. That is to say evaluating the shaded area in Fig. 5 using all **THREE** ordinates f_{-1}, f_0 and f_1. The function is represented by a 2nd order polynomial, and a 3rd order polynomial is selected to find the truncation error, namely

$$f(x) = f_0 + a x + b x^2 + c x^3 \tag{31}$$

The integral defined in Eq. 33, i.e. the area between $-h$ and 0, is obtained as,

$$\int_{-h}^{0} f(x) \, dx = \left[f_0 x + \frac{1}{2} a x^2 + \frac{1}{3} b x^3 + \frac{1}{4} c x^4 \right]_{-h}^{0}. \tag{32}$$

or upon simplification becomes,

$$\int_{-h}^{0} f(x) \, dx = f_0 h - \frac{1}{2} a h^2 + \frac{1}{3} b h^3 - \frac{1}{4} c h^4 \tag{33}$$

Fig. 5 Simpson's third rule

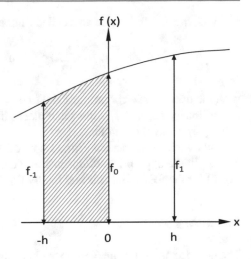

From Eq. 31, with $x = -h$

$$f_{-1} = f(x = -h) = f_0 - a h + b h^2 - c h^3 \tag{34}$$

and with $x = h$,

$$f_1 = f(x = h) = f_0 + a h + bh^2 + ch^3 \tag{35}$$

so by adding Eqs. 34–35 leading to

$$2 b h^2 = f_{-1} - 2 f_0 + f_1. \tag{36}$$

and by taking the difference of the same two equations to give,

$$2 a h = f_1 - f_{-1} - 2 c h^3. \tag{37}$$

Substituting these expressions in Eqs. 37 and 36 into Eq. 33 the integral is simplified to be,

$$I = f_0 h - \frac{1}{4} (f_1 - f_{-1}) h + \frac{1}{2} c h^4 + \frac{1}{6} (f_{-1} - 2 f_0 + f_1) - \frac{1}{4} c h^4 \tag{38}$$

and upon substitution of the limits it becomes

$$I = \frac{h}{12} (5 f_{-1} + 8 f_0 - f_1) + \frac{1}{4} c h^4 \tag{39}$$

Thus, the **5, +8, −1** rule also called Simpson's **THIRD** rule is,

$$I_{S3} = \frac{h}{12} \; (5f_{-1} + 8\,f_0 \; - f_1) \qquad\qquad (40)$$

with the truncation error proportional to $c = f'''/6$ and depending on h^4. Naturally the ordinates f_{-1} and f_1 will switch places if the area between 0 and h is required, using f_{-1} as the third ordinate.

Thus we have numerical estimates of different areas depending upon when we can apply the appropriate Simpson's rule defined in Eq. 18 for the **FIRST** rule, in Eq. 30 for the **SECOND** rule and finally Eq. 40 for the **THIRD** rule.

Summary

The need for numerical integration to obtain area- and volume-related characteristics for ships is discussed:

1. Trapezoidal, Simpson's first and second rules are derived.
2. Application of Simpson's first rule to calculate some hydrostatic properties is given for a typical ship.
3. +5, +8, −1 rule is derived and its uses are discussed.

Problems Involving Changes of Draught and Trim

7

There are a variety of practical problems involving changes of draught and trim that can be treated reasonably accurately by assuming that the changes of draught and trim are small. In the following analysis it is assumed that any changes take place within the centreline plane and, thus, do not affect the port-starboard symmetry of the vessel. Such problems might be the following:

1. The effect of adding cargo or ballast.
2. The effect of moving from freshwater to salt water.
3. The effects of docking or grounding.
4. The effects of flooding due to damage.

It is usually easiest to consider changes of mean draught and changes of trim separately.

1 The Position so far

1. To change displacement by parallel sinkage (or parallel emergence), usually denoted by T, mass must be added or removed in a vertical line through the centre of flotation (see Fig. 1).

   Changes of displacement mass are expressed in terms of tonnes per metre immersion = TPM or tonnes per centimetre immersion = TPC values which depend on waterplane area, i.e.

$$TPM = \rho A \tag{1}$$

   where A = waterplane area (m^2) and ρ = water density (tonnes/m^3).

© Springer International Publishing AG, part of Springer Nature 2018
P. A. Wilson, *Basic Naval Architecture*,
https://doi.org/10.1007/978-3-319-72805-6_7

Fig. 1 Parallel sinkage

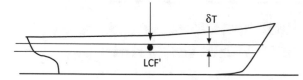

Fig. 2 Trim about LCF

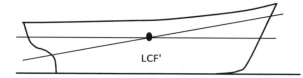

2. To change trim (usually denoted by t) at constant displacement, by moving mass
   longitudinally within the ship, the trim change takes place about LCF. That is,
   the new and old waterplanes intersect at the LCF (see Fig. 2).
   By definition,

$$\text{TRIM} = T_F - T_A = t$$

where T_F is the draught at fore perpendicular and T_A the draught at aft perpendicular.

2 Moment to Change Trim

Provided the change of trim is sufficiently small the line of action of buoyancy acts
through the longitudinal metacentre M_L vertically above the centre of buoyancy at
a point such that

$$BM_L = \frac{J_L}{\nabla}$$

where J_L is the longitudinal second moment of WPA about LCF.

To maintain the change of trim a couple must be applied to the vessel given by
(see Fig. 3)

$$\text{Moment} = \Delta \times GM_L\theta, \quad \theta \text{ in radians.}$$

Since

$$\theta = \frac{T_F - T_A}{L_{BP}} = \frac{t}{L_{BP}}$$

then

$$Moment = \frac{\Delta \ GM_L}{L_{BP}} (T_F - T_A). \tag{2}$$

Fig. 3 Moment to change trim

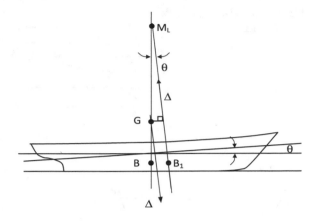

For a unit trim $(T_F - T_A = 1.0\,\text{m})$ the moment is

$$\text{MCT} = \frac{\Delta\ GM_L}{L_{BP}}$$

measured in tonne m per m if Δ is in tonnes.

In these units obviously the moment is a **mass** moment rather than a **weight** moment. MCT may be expressed in several units, with straightforward conversions, as tonne m/m, tonne m/cm, MN m/m, MN m/cm.

Clearly since moment varies linearly with trim, MCT is the additional moment required to cause a unit trim change whatever the absolute trim might be.

BM_L is of the order of ship length, whilst BG is by comparison quite small. In many cases it is sufficiently accurate to assume BM_L, GM_L and hence assume,

$$MCT \approx \frac{\Delta\ BM_L}{L_{BP}} = \frac{\rho\ \nabla}{L_{BP}}\ \frac{J_L}{\nabla} = \frac{\rho\ J_L}{L_{BP}} \tag{3}$$

3 Trimmed Draughts

To calculate the draughts T_F and T_A at the perpendiculars for a vessel trimmed by the bow (positive trim) proceed as follows:

(1) Calculate the equivalent mean draught T_E corresponding to a level keel state at the required displacement. This will be the draught of the *trimmed vessel at the LCF*.

(2) Proportion the trim between FP and AP in accordance with the distances to the perpendiculars from the LCF.

Thus for a trim $t = T_F - T_A$, using similar triangles (see Fig. 4)

$$\frac{T_F - T_E}{l_F} = \frac{t}{L_{BP}} = \frac{T_E - T_A}{l_A}.$$

Fig. 4 Trimmed draughts at perpendiculars

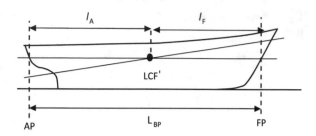

Therefore:

$$T_F = T_E + t \, \frac{l_F}{L_{BP}} \,, \qquad T_A = T_E - t \, \frac{l_A}{L_{BP}} \tag{4}$$

Note that the apportioning of trim to fore and aft perpendiculars, shown by Eq. (4), also applies to apportioning change of trim. The following sections deal with various practical situations giving rise to change of draught and trim. The changes to draught and trim as a result of flooding will be dealt with separately.

4 Adding Mass to a Vessel

The calculation of the changes of draught arising through an additional mass placed on board can be found in three stages:

1. Calculate the parallel sinkage that would occur on placing the extra mass above the LCF.
2. Calculate the change of trim arising from moving the mass from its initial position at the LCF to its final position in the vessel.
3. Calculate the final draught T_F and T_A by combining the parallel sinkage and the draught changes due to trim change.

4.1 Example 1

850 tonnes is added at 40.0 m forward of amidships of a cargo vessel of 135 m LBP. Calculate the resulting draughts at the perpendiculars from the following data:

TPC $= 22.33$ tonne/cm, MCT $= 184.6$ tonne.m/cm, LCF $= 1.66$ m aft initial draughts: $T_F = 7.3$ m, $T_A = 9.8$ m.

Fig. 5 Example of addition
of mass on a vessel

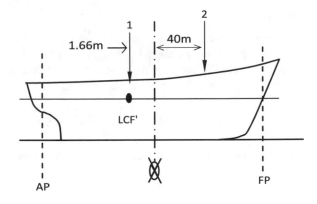

Solution

The two stages to be followed are illustrated in Fig. 5.

1. Parallel sinkage,

$$\delta T = \frac{850}{22.33} = 38.1\,\text{cm} = 0.381\,\text{m}$$

2. The trimming moment due to a movement of mass from position number 1 to position number 2 (see Fig. 5) is

$$850 \times (40 + 1.66) = 35411\,\text{tonne. m.}$$

Thus change of trim is

$$\delta t = \frac{35411}{184.6} = 191.8\,\text{cm} = 1.918\,\text{m} \qquad \text{bow down}$$

and the final draughts at the perpendiculars are obtained by adding parallel sinkage and effect of change of trim [suitably apportioned as per Eq. (4)] to the initial draughts, namely:

$$\text{Final Draught Fwd } T_F = 7.30 + 0.381 + \frac{(67.5 + 1.66)\,1.918}{135} = 8.66\,\text{m}$$

$$\text{Final Draught Aft: } T_A = 9.80 + 0.381 - \frac{(67.5 - 1.66)\,1.918}{135} = 9.25\,\text{m.}$$

4.2 Example 2

For a ship in the same initial state as example 1, sufficient mass is to be placed on board to produce a level keel draught of 9.0 m. How much mass should be added and where should its CG be placed?

Solution

The easiest solution to this problem is through obtaining each unknown quantity
from the equations for change of draught (parallel sinkage) δT and change of trim
δt, respectively. To begin with, therefore, the initial draught at LCF is required,
namely using similar triangles, or using Eq. 4:

$$T_{LCF} = 7.30 + \frac{(67.5 + 1.66)}{135}(9.80 - 7.30) = 8.581 \text{ m}.$$

Thus, the amount of parallel sinkage that took place is

$$\delta T = 9.00 - 8.581 = 0.419 \text{ m} = 41.9 \text{ cm}$$

and the required additional mass = TPC δT = 22.33 × 41.9 = 935.6 tonnes.
    Required trim change to level keel is

$$\delta t = 9.80 - 7.30 = 2.50 \text{ m} = 250 \text{ cm}. \quad \text{Bow down}$$

Corresponding Trimming Moment = 184.6 × 250 = 46150 tonne m

and

$$\text{Position of CG of Mass} = \frac{\text{Trimming Moment}}{\text{Added Mass}} = \frac{46150}{935.6} = 49.33 \text{ m} \quad \text{ahead of LCF.}$$

Hence CG of added mass = 49.33 − 1.66 = 47.67 m fwd of amidships.

5 Moving from Freshwater to Salt Water

For problems of this type it should be noted that values for TPM and MCT are
normally quoted for standard salt water (i.e. 1025 kg/m^3) as are all other hydrostatic
particulars. These are density dependent and need modification for non-standard
densities (e.g. freshwater, 1000 kg/m^3)
    Tackle this type of problem as follows:

1. Imagine the vessel moving into freshwater **without changing draught or trim**.
   In freshwater there would now be a deficit of displacement mass centred at the
   LCB
2. Calculate a parallel sinkage to restore displacement mass using a *freshwater value*
   of TPM. This additional displacement would be centred at the LCF.
3. Calculate the trim change arising from a transfer of the additional displacement
   from LCF to LCB using a *freshwater value* of MCT. This transfer keeps LCB in
   line with LCG.
4. Combine (2) and (3) to calculate new T_F, T_A.

5.1 Example 3

Consider a ship with the following properties, in salt water:

Length between perpendiculars is LBP $= 135$ m.
The ship displacement, $\Delta = 17{,}500$ tonnes at 9.0 m level keel draught.
The longitudinal centre of buoyancy LCB $= +0.61$ (fwd).
The centre of flotation, LCF $= -1.66$ m (Aft) of amidships.
Finally the moment to change trim, MCT $= 18460$ tonne m/m TPM $= 2233$.
Calculate the draughts at the perpendiculars when this ship is floating in freshwater.

Solution

- **Freshwater** displacement at the saltwater draughts (i.e. using the saltwater displacement volume) is

$$17500 \times \frac{1000}{1025} = 17073.2 \text{ tonnes.}$$

- Thus the displacement deficit $= 17500 - 17073.2 = 426.8$ tonnes.
- By linear analogy using the **freshwater** value of: TPM $= 2233 \frac{1000}{1025} = 2178.5$ tonne/m.
- Equally the **parallel sinkage** in freshwater is therefore

$$\delta T = \frac{426.8}{2178.5} = +0.196 \text{ m.}$$

- On transferring displacement deficit (or added buoyancy) from LCF to LCB (which is also LCG, **the trimming moment** $= 426.8(0.61 + 1.66) = 968.8$ tonne m.
- Then using the **freshwater** MCT $= \frac{18460}{1.025} = 18010$ tonne m/m.
- The change of trim is $\delta t = \frac{968.8}{18010} = 0.054$ m   (bow down).

The final draughts at the perpendiculars are calculated in a similar manner to example 1, namely

$$\text{Final} \quad T_F = 9.00 + 0.196 + 0.054 \frac{69.16}{135} = 9.224 \text{ m.}$$

$$\text{Final} \quad T_A = 9.00 + 0.196 - 0.054 \frac{65.84}{135} = 9.170 \text{ m.}$$

6 Docking a Vessel Trimmed by the Stern

As the water level in the dock falls below the level at which the aft end of the keel first touches a keel block, the vessel will pivot about the after end until the keel touches all the blocks along the keel. The load on the keel increases as the level falls and reaches a maximum just before the keel touches or **sues** all along. The local hull structure must be strong enough to sustain the maximum load, and the vessel must remain stable until the vessel settles onto the blocks, as **shores** cannot be placed until she stops moving.

The problem is to evaluate the keel load P for any given fall in water level. The starting point is that, since the point of contact at the keel remains fixed in space, the change of draught at the point of contact is equal to the fall in water level from that at which contact is first made (see Fig. 6).

6.1 Example 4

The cargo vessel of previous examples, LBP=135m, enters a drydock with a mean draught of 5.0 m and a trim by the stern of 4.0 m. At the given draught the following data applies:

$$TPM = 2027, MCT = 14190 \text{ tonne.m/m}, LCF = 0.65 \text{ m} \text{ fwd of amidships.}$$

Neglecting any changes in the above data, and given that the aftermost keel block is 8.0 m ahead of AP, calculate the keel load, mean draught and trim,
   (a) after the water level has fallen 1.0 m and
   (b) as the keel is about to ground all along.

Solution

Initial draught at LCF and the aft keel block are required for (a) and (b). These are obtained using relationships for appropriate similar triangles [i.e similar to Eq. (4)].

Initial draught at aft keel block $= 5.00 + \frac{4.0\,(67.5-8.0)}{135} = 6.763$ m. $=$ depth of water above keel block at first contact.

Fig. 6 Ship being brought into dry dock

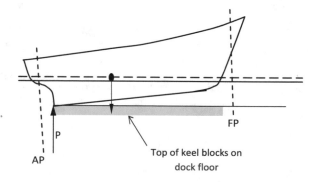

P

FP

Top of keel blocks on
dock floor

AP

Initial draught at LCF $= 5.00 - \frac{4.0 \times 0.65}{135} = 4.980\,\mathrm{m}$

1. 1.0-m fall in water level: **Parallel** rise due to the effect of keel block load P:

$$\delta T = \frac{P}{2027}\,\mathrm{m}$$

Trimming moment due to keel block load transfer from LCF to point of contact (i.e. aft keel block):
Trimming moment $= P(67.5 + 0.65 - 8.0) = 60.15P$ tonne. m and resultant change of trim, δt,

$$\delta t = \frac{60.15P}{14190} \quad \text{bow down.}$$

Change of draught at point of contact (i.e. aft keel block) is

$$\delta T_c = -\frac{P}{2027} - \delta t\,\frac{60.15}{135} = -P\left\{\frac{1}{2027} + \frac{60.15^2}{14190 \times 135}\right\}$$

$$\delta T_c = -0.002382P.$$

$\delta T_c = -1\,\mathrm{m}$ as a result of fall of water level by 1 m. Thus the keel load is

$$P = \frac{1.0}{0.002382} = 419.8\,\text{tonnes}$$

Note that the keel load was obtained in units of mass rather than force; therefore,

$$P = 9.81 \times 419.8 = 4118\,\mathrm{kN} = 4.118\,\mathrm{MN}.$$

Now that the value of the keel load (in tonnes) is known, parallel rise (or emergence) and change of trim due to 1-m fall of the water level can be obtained as
Parallel rise

$$\delta T = \frac{419.8}{2027} = 0.207\,\mathrm{m}.$$

Change of trim

$$\delta t = \frac{60.15 \times 419.8}{14190} = 1.780\,\mathrm{m} \quad \text{(bow down)}$$

Draughts after 1.0-m fall in water level, in a similar manner to examples 1 and 3 are

$$T_F = 3.00 - 0.207 + \frac{1.780\,(67.5 - 0.65)}{135} = 3.67\,\mathrm{m}.$$

$$T_A = 7.00 - 0.207 - \frac{1.780\,(67.5 + 0.65)}{135} = 5.89\,\mathrm{m}.$$

2. Ship to level keel case to touch all blocks. Considering the initial condition of
   the ship, with trim by the stern (or stern down), a change of trim bow down of
   $t = 4$ m is required to level keel. Thus, the trimming moment to change trim is:
   $\delta t = 4.0$ m
   $MCT = 4.0 \times 14190 = 56760$ tonne.m $= 60.15P$
   As this trimming moment is equal to the moment due to load from the keel being
   transfered from LCF to aft keel block. Thus, keel load at just *about to touch* all
   along is

$$P = \frac{56760}{60.15} = 943.6 \text{ tonnes} \quad or \quad 9.257 \text{ MN}$$

In order to find the corresponding level keel draught, only the parallel rise is
required to be used in conjunction with the initial draught at LCF. Therefore,
parallel rise is

$$\delta T = \frac{943.6}{2027} = 0.466 \text{ m}$$

and final level keel draught at LCF (and everywhere else of course) is $T = 4.980 - 0.466 = 4.514$ m.
Consequently fall of water level is $6.763 - 4.514 = 2.248$ m, using the initial
draught at the keel block.

7 Variation of Hydrostatic Particulars with Draught

It should be noted that values of Δ, LCB, LCF, TPM and MCT vary as draught
changes, typical level keel information being as follows in Table 1, positive forward
of amidships:

Where large changes of draught and trim occur it may be necessary to do the calculation twice: firstly with an assumed mean draught (and corresponding particulars)
and secondly using particulars for a draught close to the final answer.

Table 1 Variation of hydrostatics with waterline

T (m)	Δ (tonnes)	LCB (m)	LCF (m)	TPM	MCT (tonne m/m)
7.0	13181	+1.167	−0.517	2132	16208
5.0	9020	+1.676	+0.646	2027	14190
3.0	5079	+2.134	+1.596	1908	12344

8   The Inclining Experiment

It is usual to carry out an inclining experiment on a new ship when the vessel is as near to completion as can be arranged. Such an experiment is repeated after every major refit during the life of the ship.

8.1   Purpose

The purpose of the experiment is to determine the lightship mass together with the longitudinal and vertical positions of the Centre of Gravity of the vessel in the lightship state. In the lightship state the vessel is empty, i.e. it has no fuel, cargo, water ballast, stores, passengers or crew on board.

The mass and Centre of Gravity of the vessel in any subsequent state of loading can be found by adding the mass and moments of mass of each load item to the basic lightship values. A knowledge of the vertical position of C of G is, of course, crucial to the assessment of stability.

At the design stage lightship mass and C of G are estimated by comparison with data from previously completed similar ships. Later on, direct calculations based on manufacturers weight data for equipment and from records of weight added or removed during building are used to improve these estimates. The inclining experiment provides the final, definitive, values of mass and C of G and will, hopefully, confirm the accuracy of the earlier estimates.

8.2   Method

Hydrostatic data computed with data from the ships lines plan are taken as accurate - barring accidental blunders the data should be within $\frac{1}{4}\%$ or so of the correct values, anyway. Ship mass and LCG, as inclined, are found from measurements of draught and trim, together with a measurement of water density taken at the time of the experiment, using hydrostatic data. VCG, as inclined, is found from the computed height of the transverse metacentre using a value of GM_T determined in the experiment.

Ballast weights large enough to heel the vessel $2°$–$3°$ to either side are moved in groups a preset distance across the deck from centreline to port and to starboard (see Fig. 7). The ballast will be moved several times, and after each movement the heel angle will be measured using a long pendulum. Any swinging of the pendulum should be damped by immersing the pendulum bob in oil or water. Since the heel angles are small it is sufficiently accurate to write

$$W \times D = \Delta \times GM_T \times \tan\phi = \Delta \times GM_T \frac{d}{l}$$

on moving a weight W a distance D across the deck from a condition in which the ship is upright.

Fig. 7 Inclined section

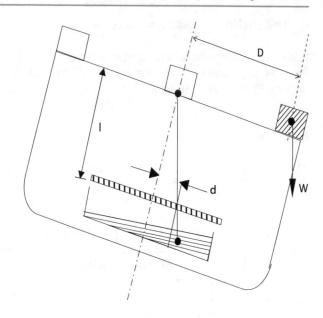

8.3 Precautions to Observe

In order to ensure that the estimates of lightship mass, LCG and VCG are accurate the following precautions should be observed during an inclining experiment.

- The ship should be as near to completion as can be arranged.
- The experiment should be conducted in slack water, that is at high or low tide, and in calm weather.
- All moveable items, including people, that can be put ashore should be removed from the ship.
- The ship should be inspected and careful note made of (i) items not yet on board and (ii) contractor's gear and scrap materials to be removed before completion.
- Bilge spaces should be pumped dry. The state of all tanks containing liquids should be inspected so that appropriate free surface corrections may be allowed. Ideally all tanks (fuel oil, lubricating oil, engine sumps, boilers, freshwater tanks, sludge tanks, ballast tanks, etc.) should be either pumped dry or pressed full.
- Moorings lines should be slack, and the vessel should float freely away from the dockwall and preferably head to wind. There should be no floating stages moored alongside.
- The actual weights of the moveable ballast used to incline the ship should be checked, as also should the length of the pendulum and its method of hanging.

8.4   Measurements of Draught

Measurements of draught from draught marks on ship side or, in the case of small yachts, from measurements of freeboard should be made at several points round the vessel (both port and starboard sides). A transparent sighting tube should be used to obtain a steady water level (inside the tube) regardless of small waves present on the water surface (see Fig. 8). Water samples should also be taken at several points round the vessel to establish water density.

8.5   Corrections to Lightship

Once the inclining experiment has been completed values of ship mass, LCG and VCG for the ship as inclined must be corrected to equivalent lightship values by accounting item by item for things to be added to or removed from the ship to reach the lightship state. One item to remove will be the weights used to incline the ship. An example calculation is shown in Table 2:

Fig. 8 Draught measurements

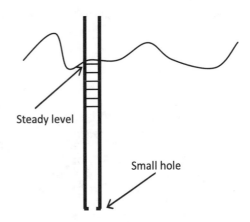

Steady level

Small hole

Table 2 Corrections to lightship displacement

Item	Mass	VCG	Vert moment	LCG	Long moment
Ship as inclined	4155	9.3	38642	−8.2	−34071
To remove inclining weights	−102	15.7	−1601	+2.5	−255
To remove contractor's gear	−53	12.1	−641	−71.0	+3763
To add radar scanner	+2	30.5	61	−57.4	−115
To add lifeboats	+15	23.0	345	−67.0	−1005
To add anchors and cables	36	8.5	306	74.5	2682
Totals (lightship)	4053		37112		−29001

$$\text{LIGHTSHIP } VCG = \frac{37112}{4053} = 9.16\,\text{m}, LCG = \frac{-29,001}{4053} = -7.16\,\text{m}.$$

Summary

1. TPM, TPC for evaluating small changes in draught and MCT for evaluating small changes in trim are defined.
2. Various examples are provided to illustrate changes in draught and trim.
3. A parallel sinkage (or emergence) is due to mass added (or subtracted) or displacement deficit (or surplus) applied at LCF.
4. Change of trim (bow or stern down) is due to moving the mass (added or subtracted) or displacement deficit or surplus from LCF to its appropriate position along the ship
5. The inclining experiment is outlined.
6. The aims of the inclining experiment are defined, namely obtaining lightship mass and lightship LCG and VCG.
7. The various measurements taken during an inclining experiment are outlined.
8. The precautions required before the inclining experiment are summarised.
9. An example of subsequent corrections to lightship mass LCG and VCG is given.

Transverse Initial Stability Topics

8

1 Righting and Heeling Moments at Small Angles

At this stage transverse stability (i.e. stability in heel/roll) will be treated for angles sufficiently small that the line of action of the force of buoyancy acts through the transverse metacentre as a fixed point. Angles of heel may be induced by the action of wind or by moving mass within the vessel or by similar causes. These causes result in a heeling moment on the vessel that, at the equilibrium heel angle, is counteracted by a couple formed from the weight and buoyancy forces.

For small angles of heel (up to about 7°) the weight (or buoyancy) **roll restoring moment** is (see Fig. 1),

$$M_1(\varphi) = \Delta GZ = \Delta GM_T \sin(\varphi) \tag{1}$$

where G is the **Centre of Gravity** and M_T is the **transverse metacentre**.

Transferring a mass of weight w across the vessel perpendicular to centreline plane through a distance d (from A to B, as shown in Fig. 2) will introduce a **heeling moment** due to the shift of lines of action of w. The heeling moment in the final position of the vessel is

$$M_2(\varphi) = wd \cos(\varphi). \tag{2}$$

Equilibrium requires the roll restoring (or righting moment) to equal the heeling moment, i.e.

$$M_1(\varphi) = M_2(\varphi)$$

leading to,

$$wd\cos(\varphi) = \Delta GM_T sin(\varphi)$$

© Springer International Publishing AG, part of Springer Nature 2018
P. A. Wilson, *Basic Naval Architecture*,
https://doi.org/10.1007/978-3-319-72805-6_8

Fig. 1 Roll restoring
moment

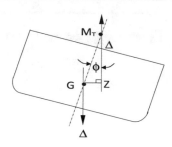

Fig. 2 Moment of inclining
masses

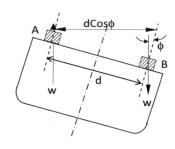

Fig. 3 Position of B, G and
M_T

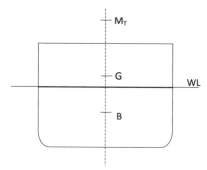

or

$$tan\phi = \frac{wd}{\Delta\, GM_T} \qquad (3)$$

Although not explicitly stated, the above analysis assumes that the Centre of
Gravity G is as it was with the mass in its original position (A) and the displacement
weight Δ is inclusive of w. Obviously movement of mass moves the Centre of
Gravity to a new position, and the vessel heels until the new Centre of Gravity is
in line vertically with the new centre of buoyancy and hence with the transverse
metacentre M_T. Check for yourself that calculating by this second method leads to
the same answer as given above. Calculation of GM_T is done as follows (see Fig. 3):

$$GM_T = KM_T - KG = KB + BM_T - KG \qquad (4)$$

where KG is height of C of G above keel datum, KB is height of centre of buoyancy above keel and,

$$BM_T = \frac{J_T}{\nabla}$$

is height of transverse metacentre above the centre of buoyancy.

Clearly as mass is added to the vessel (e.g. by placing cargo on board), draught will change, displacement weight and volume will change (leading to changes of KB and BM_T) and the combined C of G, inclusive of the additional mass, will also change. These effects will combine to produce the final change in GM_T. The easiest way to view the effect of an addition of mass is to consider it placed in two stages:

1. Place on board above G so as to avoid heeling the vessel (see Chap. 7). Calculate new values for Δ, KB, BM_T, KG and hence GM_T.
2. Shift to the final position and calculate the heel angle produced using Eq. (3).

The easiest way to view the movement of mass inside the vessel is to consider it in two stages:

1. Move the mass to its final height above keel by a vertical movement that will not cause heeling. Calculate a new KG and hence a new GM_T.
2. Move the mass across the ship to its final position and calculate the heel angle so caused, using Eq. (3).

2 Metacentric Height Diagram for a Rectangular Box

For a solid uniform rectangular block (see Fig. 4):

$$KB = \frac{T}{2} \quad \text{(i.e. half-draught)}$$

Fig. 4 Calculation of M_T for a rectangular box

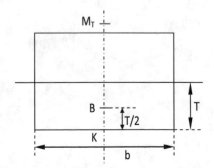

$$J_T = \frac{Lb^3}{12} \quad (L = \text{length of block})$$

and, thus,

$$BM_T = \frac{J_T}{\nabla} = \frac{\frac{1}{12} Lb^3}{LbT} = \frac{b^2}{12\,T}$$

Hence,

$$KM_T = \frac{T}{2} + \frac{b^2}{12\,T} \tag{5}$$

$$\frac{d\,(KM_T)}{dT} = \frac{1}{2} - \frac{b^2}{12\,T^2}$$

and KM_T has a minimum value when,

$$\frac{d(KM_T)}{dT} = 0$$

that is, when

$$\frac{b}{T} = \sqrt{6} = 2.45$$

Equation (5) can be put in the following form:

$$\frac{KM_T}{b} = 0.5\,\frac{T}{b} + \frac{b}{12\,T}$$

The metacentric diagram for the block is as shown in Fig. 5.
Most ship forms exhibit a similar curve with a minimum at,

$$\frac{b}{T} \approx 2.5$$

Fig. 5 Variation of box
parameters with KM_T

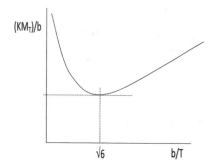

3   Stability of a Uniform Square Sectioned Log

3.1   Log Floating with One Face Horizontal

A totally solid uniform log is not, normally, stable floating with one face horizontal, as sketched in Fig. 6. In this case the log is unstable over a range of draughts for which M_T is below G. These draughts, shown schematically in Fig. 6, can be obtained using Eq. (4) and setting $GM_T = 0$. KM_T is already known from the previous section [Eq. (5)] and $KG = d/2$, where d is the side length of the square log. Accordingly the limit draughts can be evaluated from:

$$6\,T^2 \; - \; 6d\,T \; + \; d^2 \; = \; 0$$

as $0.2113d$ and $0.7887d$. Stable and unstable draught ranges can then be identified by finding which draughts result in positive and negative values of GM_T using Eqs. 4 and 5. You are advised to obtain the above equation and identify regions of stability and instability.

3.2   Log Floating with One Diagonal Horizontal

In this case identification of stable and unstable draughts has to be carried out for two subcases, namely when the uniform solid log is floating at a draught,

1. Below the diagonal, as shown in Fig. 7, and
2. Above the diagonal, as shown in Fig. 8.

Fig. 6 Stability range for a square sectioned log

Fig. 7 Square sectioned log floating on diagonal with shallow waterline

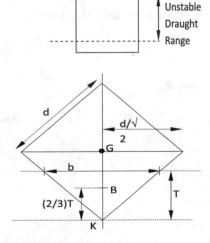

Fig. 8 Square sectioned log
floating on diagonal with
deep waterline

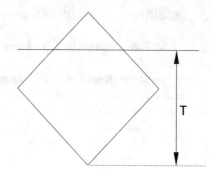

Consider the first subcase for a log of side length d floating at a draught T,

$$T < \frac{d}{\sqrt{2}}$$

Let $b =$ waterline breadth at draught T, as shown in Fig. 7. Then,

$$\nabla = \frac{LbT}{2}$$

$$J_T = \frac{Lb^3}{12}, \quad BM_T = \frac{J_T}{\nabla} = \frac{b^2}{6\,T}$$

and as from the triangular shape,

$$b = 2\,T$$

then

$$BM_T = \frac{2}{3}T$$

From geometrical consideration,

$$KB = \frac{2}{3}T$$

and, thus,

$$KM_T = \frac{4}{3}T$$

Now as,

$$KG = \frac{d}{\sqrt{2}}$$

then,

$$GM_T = \frac{4}{3}T - \frac{d}{\sqrt{2}} > 0 \quad \text{if log is stable.}$$

Thus, log is stable in this case if,

$$T > \frac{3}{4\sqrt{2}} d$$

Now, by definition,

$$\text{Specific Gravity} = \frac{\text{Material Density}}{\text{Fresh Water Density}}$$

Hence, for a log floating in equilibrium,

$$\text{Specific Gravity} = s = \frac{\nabla}{\text{Block Volume}}$$

or

$$s = \frac{\frac{LbT}{2}}{L\,d^2} = \left(\frac{T}{d}\right)^2$$

The minimum specific gravity for subcase (1) to be stable is, thus,

$$s = \left(\frac{3}{4\sqrt{2}}\right)^2 = \frac{9}{32} = 0.28125$$

It is also possible to consider subcase (2) where

$$T > \frac{d}{\sqrt{2}}$$

(see Fig. 8). This is a more complicated subcase, in terms of obtaining KB and BM_T. In this case it can be found that for the log to be stable,

$$T < \frac{5}{4\sqrt{2}} d$$

This corresponds to a maximum specific gravity $s = 1 - 0.28125 = 0.71875$ for which subcase (b) is stable. The Ranges of Stability and instability are illustrated in the sketch of Fig. 9.

Fig. 9 Stability range of square sectioned log floating on a diagonal waterline

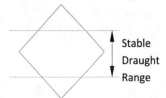

4 Morrish's Formula for KB

At an early stage in the ship design process it is necessary to ensure that the ship will have an adequate GM_T for all reasonable loading conditions. This requires the estimation of KB and BM_T. The following method, due to Morrish, (S.W.F. Morrish TRINA, 1892) gives a reasonably accurate quick estimate of KB.

The basis is to represent the **vertical distribution of waterplane area** by a simple quadrilateral, as shown in Fig. 10, where

$$OK = T = \text{Load Draught}$$

$$OA = A = \text{Load Waterplane Area} = C_W \, L \, B$$

$$OC = \frac{\nabla}{T} = \text{Mean Waterplane Area} = KD$$

$$CE = \frac{\nabla}{A} = \text{Mean Draught}$$

Triangle AEK is a simple approximation to the waterplane area variation with distance from baseline satisfying the requirement that,

$$\nabla = \int_0^T A(z)dz$$

since

$$\text{Area} \ \ OCDK \ = \ \frac{\nabla}{T} T \ = \ \nabla$$

$$\text{Area} \ \ KED \ = \ \frac{1}{2} \frac{\nabla}{T} \left(T - \frac{\nabla}{A} \right) = \frac{1}{2} \nabla \left(1 - \frac{\nabla}{A \, T} \right)$$

Fig. 10 Approximation of waterplane area using linear variations used in Morrish's method

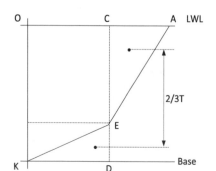

$$\text{Area} \quad AEC = \frac{1}{2}\frac{\nabla}{A}\left(A - \frac{\nabla}{T}\right) = \frac{1}{2}\nabla\left(1 - \frac{\nabla}{A\,T}\right) = \text{Area } KED$$

Thus,

Area $OAEK$ = Area $OCDK$ + Area AEC − Area KED = ∇ (as required).

The height of the centroid of $OAEK$ corresponds to KB and is found by taking moments of area above the base, as follows:

$$\nabla\, KB = \nabla\frac{T}{2} - \frac{1}{2}\nabla\left(1 - \frac{\nabla}{A\,T}\right)\frac{ED}{3} + \frac{1}{2}\nabla\left(1 - \frac{\nabla}{A\,T}\right)\left(T - \frac{EC}{3}\right)$$

$$= \nabla\frac{T}{2} + \frac{\nabla}{2}\left(1 - \frac{\nabla}{A\,T}\right)\frac{2\,T}{3}$$

Thus,

$$\frac{KB}{T} = \frac{1}{2} + \frac{1}{3}\left(1 - \frac{\nabla}{A\,T}\right) = \frac{1}{3}\left[2.5 - \frac{\nabla}{A\,T}\right]$$

But

$$\frac{\nabla}{AT} = \frac{LBTC_B}{LBC_W T} = \frac{C_B}{C_W}$$

and, thus,

$$\frac{KB}{T} = \frac{1}{3}\left[2.5 - \frac{C_B}{C_W}\right] \qquad (6)$$

Thus Eq. 6 is the so-called **Morrish** formulae for estimating KB.

Note that Morrish's formula was originally devised in terms of the distance of B below the waterline, for which,

$$\frac{OB}{T} = 1 - \frac{KB}{T} = \frac{1}{3}\left[0.5 + \frac{C_B}{C_W}\right]$$

These formulae are usually within 1 or 2% of the correct answer for normal hull forms.

5 Munro-Smith Estimate of BM_T

As considered earlier, in Chap. 5, for a simple waterplane curve of the form,

$$y = \frac{B}{2}\left(1 - \left(\frac{2x}{L}\right)^n\right)$$

The waterplane area is

$$A = L B \left(1 - \frac{1}{n+1}\right) = L B C_W$$

Hence,

$$n + 1 = \frac{1}{1 - C_W}$$

The transverse second moment of area is

$$J_T = \frac{L B^3}{12} \left[1 - \frac{3}{n+1} + \frac{3}{2n+1} - \frac{1}{3n+1}\right]$$

Now,

$$2n + 1 = 2(n+1) - 1 = \frac{2}{1 - C_W} - 1 = \frac{1 + C_W}{1 - C_W}$$

and,

$$3n + 1 = 3(n+1) - 2 = \frac{1 + 2 C_W}{1 - C_W}$$

Hence,

$$J_T = \frac{L B^3}{12} \left[1 - 3(1 - C_W) + \frac{3(1 - C_W)}{1 + C_W} - \frac{(1 - C_W)}{1 + 2 C_W}\right]$$

$$J_T = \frac{L B^3}{12} \left[3 C_W - 2 + (1 - C_W)\frac{(2 + 5 C_W)}{(1 + C_W)(1 + 2 C_W)}\right]$$

$$J_T = \frac{L B^3}{12} \frac{(3 C_W - 2)(1 + 3 C_W + 2 C_W^2) + (2 + 3 C_W - 5 C_W^2)}{(1 + C_W)(1 + 2 C_W)}$$

$$J_T = \frac{L B^3}{12} \frac{6 C_W^3}{(1 + C_W)(1 + 2 C_W)}$$

$$BM_T = \frac{J_T}{\nabla} = \frac{J_T}{L B T C_B} = \frac{B^2}{T} \frac{C_W^3}{2 C_B (1 + C_W)(1 + 2 C_W)} \tag{7}$$

Eq. 7 is the so-called **Munro-Smith** formula for estimating BM_T.

The benefit of this formula and of Morrish's formula is that quite good estimates of BM_T and KB can be made even before a full lines plan is drawn, so long as suitable values of block coefficient (C_B) and waterplane area coefficient (C_W) are available. At a later stage in the design more accurate values will be obtained by numerical integration (e.g. Simpson's rules) from the lines plan, as illustrated in Chap. 6.

6   Initial Estimate of Ship Moulded Beam

The height of the transverse metacentre above the keel is closely dependent on the choice of ship beam. At the initial stage of ship design the choice of beam is on the basis of a choice of GM_T thought likely to give adequate transverse stability.

As an example, consider a car ferry for which considerations of likely values of KG and GM_T suggest that $KM_T = 11.5$ m is a suitable value, assuming the following particulars:

$$\nabla = 12,360 \, \text{m}^3, LBP = 150 \, \text{m}, C_B = 0.576, C_W = 0.737$$

estimates of beam and draught are required.

$$B \, T = \frac{\nabla}{L \, C_B} = \frac{12360}{150 \times 0.576} = 143.1 \, \text{m}^2$$

Using Morrish's formula, i.e. Eq. (6):

$$\frac{KB}{T} = \frac{1}{3}\left(2.5 - \frac{0.576}{0.737}\right) = 0.573$$

Using Munro-Smith formula, i.e. Eq. (7):

$$BM_T = \frac{B^2}{T} \frac{0.737^3}{2 \times 0.576 \, (1 + \, 0.737)(1 + 2 \ 0.737)}$$

$$BM_T = 0.08086 \, \frac{B^2}{T}$$

The following table can then be constructed using suitably selected trial values for the beam B. The term suitably refers to selecting values for B which will be relatively close to the final answer. Otherwise the table may become unnecessarily long (Table 1).

Clearly, B $= 24.0$ m and T $= 5.96$ m are suitable values, as corresponding KM_T is very close to the required value. In other cases further iteration between two trial values may be necessary. Please note that there is another set of beam and draught values which also satisfy the above requirements. However, the corresponding draught is rather large and, hence, may be unpractical.

Can you find this set other of values?

Table 1   Trial and error estimation for moulded beam

Trial B (m)	Corresponding T (m)	KB (m)	BM_T (m)	KM_T (m)
22.0	6.51	3.73	6.20	9.93
24.0	5.96	3.42	8.05	11.47
26.0	5.50	3.15	10.24	13.39

7   Losses of Transverse Stability - *Virtual* Centre Gravity Problems

There are a number of circumstances which reduce the transverse stability of a floating body, each of which can be analysed by calculating a virtual change to the body KG and as a consequence to arrive at a virtual loss of GM_T. The body becomes unstable if this virtual loss exceeds the initial GM_T value. Three typical problems are:

1. The effect of suspended weights free to swing,
2. The effect of liquid free surfaces in tank spaces,
3. The loss of stability due to docking or grounding.

7.1   Suspended Weights

If a suspended weight w is free to swing on a wire (see Fig. 11), then the line of action of weight will be along the wire from the point of suspension G_2 at all angles of heel. G_2 becomes the **Virtual Centre of Gravity** of the weight. Raising its Centre of Gravity from its initial, true, position at G_1 to its virtual position G_2 will effectively raise the overall ship KG and hence reduce GM_T.

Clearly,

$$\delta KG = \frac{w \ G_1 G_2}{\Delta}$$

is the effective loss in GM_T.

For a weight suspended from a derrick the point of suspension is at A if the derrick is prevented from swinging by suitable **guys** or vangs and at B if the whole assembly is free to swing (see Fig. 12).

Note that the full loss of stability occurs at the instant when the weight is raised off its stowage and becomes free to swing. If an item of cargo is swung overside on a derrick whose movement is fully controlled the heel angle produced is found as (see Fig. 13)

Fig. 11 Weights suspended by a crane

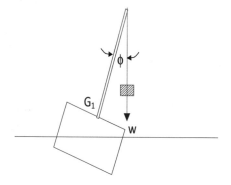

Fig. 12 A derrick under control booms or vangs

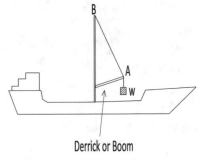

Fig. 13 Derrick under full control

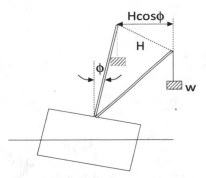

$$tan\varphi = \frac{w \ H}{\Delta \ GM'_T}$$

where GM'_T is the effective transverse GM_T with the virtual CG of the cargo raised to the point of suspension provided the vessel is upright before the cargo is moved.

7.2 Liquid Free Surfaces

If a liquid in a tank has a free surface and is free to flow as the vessel heels, then the Centre of Gravity of the liquid will move as the liquid flows, say from G_o (upright) to G_1 (heeled), as shown in Fig. 14. The lines of action of liquid weight upright through G_0 and heeled through G_1 will intersect at G_2. Thus, the Centre of Gravity of the liquid is effectively raised from its true initial position G_o to a virtual position G_2. This will effectively raise the overall vessel KG and hence reduce the effective GM_T. There is a clear analogy between the transfer of wedges of buoyancy, transverse movement of centre of buoyancy and BM_T, discussed in Sect. 5.8, and the above transfer of liquid on heeling, giving an analogy between BM_T and G_oG_2.

Thus,

$$G_o \, G_2 = \frac{j_T}{\nabla_1}$$

Fig. 14 Free surface effects

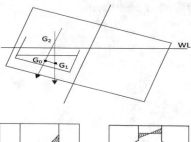

Fig. 15 Reduction of free
surface area in a vertical
hopper

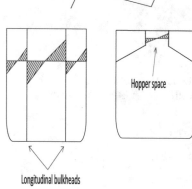

Longitudinal bulkheads

where j_T is the transverse second moment of area of the liquid-free surface and ∇_1 is
the volume of liquid in the tank. The effective increase of mass moment above base is

$$\Delta \ \delta KG = \rho_L \ \nabla_1 \ G_o \, G_2 = \rho_L \ j_T$$

$$\delta KG = \frac{\rho_L \ j_T}{\Delta} = \frac{\rho_L}{\rho_s} \frac{j_T}{\nabla} \qquad (8)$$

where ρ_L = density of tank liquid and ρ_s = density of water in which vessel floats
(usually salt).

Note j_T is measured about a fore-and-aft axis through the centroid of the liquid
free surface. The actual position of the tank in the vessel is not important. There may
be several tanks in the vessel which potentially have free surfaces. The effects are, of
course, cumulative. Large tanks are frequently divided longitudinally by liquid-tight
bulkheads to reduce free surface losses (see Fig. 15). Cargo tanks in vessels carrying
chemical products may be fitted with hopper spaces to achieve the same result (see
Fig. 15). The reduction in the increase KG (or decrease GM_T) is achieved through
decreasing j_T. For example in an undivided tank, with rectangular free surface, (see
Fig. 16)

$$j_T = \frac{lb^3}{12}$$

Fig. 16 Subdivision
longitudinally in cargo tanks

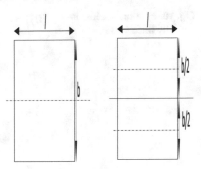

Fig. 17 Change of G due to
replacement of sea water for
oil in fuel tanks

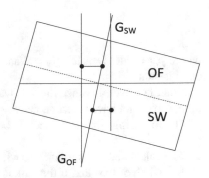

whilst in the same tank with one central division:

$$j_T = \frac{1}{12} \, l \left(\frac{b}{2}\right)^3$$

per tank resulting in

$$2 \frac{1}{12} l \left(\frac{b}{2}\right)^3 = \frac{1}{4} \frac{l \, b^3}{12}$$

total value for j_T.

Thus, for a rectangular free surface a single central bulkhead will reduce the free surface loss to 25% of that for an undivided tank. Dividing the tank space in three will further reduce the loss.

To reduce free surface losses in submarine fuel tanks the tanks are topped up with sea water as fuel is used. The oil fuel (OF) floats on the sea water (SW). Fuel for the engines is, of course, drawn from the top of the tank. The interface between the fuel and water remains level as the vessel heels. The movement of fuel and water is such that the virtual CG of the fuel is lowered to G_{OF} and that of the sea water raised to G_{SW} (see Fig. 17). The net effect is a virtual rise in the submarine KG given by,

$$\delta KG = \left(\frac{\rho_w}{\rho_s} - \frac{\rho_{of}}{\rho_s}\right) \frac{j_T}{\nabla}$$

Fig. 18 Free surface loss in
partially filled tanks

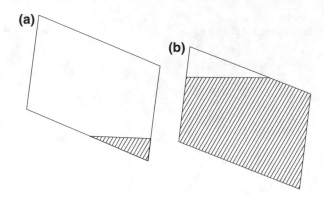

where ρ_{of} = Fuel Density, ρ_w = Water Density inside Tank and ρ_s = Water Density outside submarine.

Note, however, because the salt water (SW) is taken on board over a period of time, it may not have the same density as the water currently outside the submarine.

The following comments need to be noted:

1. The calculated free surface loss will normally assume the free surface extends the full width of the tank. If the tank is nearly empty (as shown in Fig.18a) or nearly full (as shown in Fig. 18) the full loss may only apply for small heel angles.
2. The effects of waves generated inside the tank as a result of ship motions in rough weather result in dynamic effects which will usually act to reduce roll motion. This is beneficial to the vessel, so that the presence of a free surface is not harmful unless it reduces effective GM_T below a minimum safe level.

7.3   Stability Losses Due to Grounding or Docking

When a vessel is docked in a drydock it will ground on the docking blocks as the dock water level falls. The difference between the buoyancy available at the current water level and the ship weight is manifest as a reaction load between the ship and the keel blocks. Inadvertently grounding on a rock has the same result. This is illustrated in Fig. 19, where

Δ_0 = Initial free floating displacement before grounding.

Δ_1 = Displacement at current draught/trim.

M = Transverse metacentre at current draught and trim.

$P = \Delta_0 - \Delta_1$ = Grounding load at keel.

Fig. 19 Stability loss due to
grounding or docking

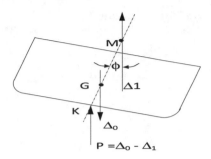

Taking moments about G the righting moment at a small angle of heel φ is given by

$$\Delta_1 GM \sin(\varphi) - PKG \sin(\varphi).$$

This moment can be expressed in terms of an effective GM_T based either on the current displacement Δ_1 as,

$$\Delta_1 GM_1 \sin(\varphi)$$

where

$$GM_1 = GM - \frac{PKG}{\Delta_1}$$

or alternatively, based on the initial displacement Δ_o as,

$$\Delta_0 GM_0 \sin(\varphi)$$

This is achieved by rewriting the righting moment as,

$$(\Delta_0 - P)GM \sin(\varphi) - PKG \sin(\varphi) = \Delta_0 GM \sin(\varphi) - PKM \sin(\varphi)$$

leading to,

$$GM_0 = GM - \frac{PKM}{\Delta_o}$$

Either way, the effect of the keel load is to reduce the effective GM of the vessel which will become unstable if GM_1 or GM_o falls below zero.

Docking shores must be in place to support the vessel whilst the vessel remains stable (see Fig. 20). This can be important when a vessel enters dock heavily trimmed by the stern, because shores cannot be placed until the keel grounds all along the blocks.

Fig. 20 Shores used in dry
docking

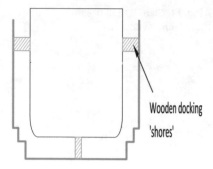

Wooden docking
'shores'

8 Summary

1. Heeling and righting (for small angles) moments are defined; righting lever GZ
   (for small angles of heel) is defined.
2. Calculation of $GM_T = KB + BM_T - KG$, vertical distance of transverse metacentre above the Centre of Gravity is illustrated.
3. Stability of a uniform solid square log is discussed.
4. Morrish's and Munro-Smith's formulae for estimating KB and BM_T are derived,
   and their use in preliminary design illustrated.
5. Various factors adversely affecting GM_T are discussed:

   a. Suspended weights reduce GM_T.
   b. Liquid-free surface in tanks reduces GM_T and is one of the most important
      factors contributing to loss of GM_T.
   c. Grounding and docking reduce GM_T

Wall-Sided Formula and Applications

9

So far transverse stability has been considered only for heel angles up to, say 5°–7° for which it can be reasonably assumed that the line of action of buoyancy acts through the transverse metacentre M_T.

Before moving on to a general treatment of large angle stability it is worth considering a simple formula valid for any range of heel angle within which the sections near the waterline remain vertical, or nearly so. In many cases such a formula is reasonably accurate up to the point at which the deck edge immerses. The **wall-sided formula** allows for movement of the centre of buoyancy parallel to centreline as well as parallel to the upright waterline.

1 Wall-Sided Formula

Consider a **slice** of the vessel between x and $x + \delta x$, whose section is as sketched in Fig. 1,
where
  $W_o L_o$ = Upright waterline,
  $W_1 L_1$ = Heeled waterline,
  φ = Heel angle,
  $\oplus$ = Centroids of immersing and emerging wedges,
  y = Waterline half-breadth.
  As the vessel heels,
  Volume of immersing and emerging wedges = $\delta \nabla = 0.5 y^2 \tan \varphi \, \delta x$.

Due to the transfer of volume from upgoing to downgoing side there is an elemental change of moment of volume about centreline given by,

$$\delta M_H = \frac{4}{3} y \, \delta \, \nabla = \frac{2}{3} y^3 \, \delta x \, \tan \varphi.$$

© Springer International Publishing AG, part of Springer Nature 2018
P. A. Wilson, *Basic Naval Architecture*,
https://doi.org/10.1007/978-3-319-72805-6_9

Fig. 1 Heeled ship section
for wall-sided calculation

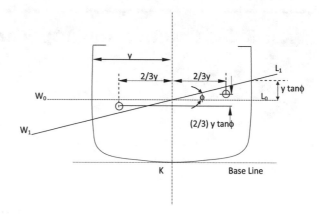

Similarly, due to the movement of $\delta\nabla$ away from the baseline there is a change
of moment of volume above the keel given by,

$$\delta M_V = \frac{2}{3} \, y \, \tan\varphi \, \delta\nabla = \frac{1}{2}\frac{2}{3} y^3 \, \delta x \, \tan^2\varphi.$$

These elemental changes of moment can be integrated along the ship length to
give total moment changes for the whole ship of:

$$M_H = \frac{2}{3} \int\limits_{x_A}^{x_F} y^3 \, dx \, \tan\varphi$$

and,

$$M_V = \frac{1}{2}\frac{2}{3} \int\limits_{x_A}^{x_F} y^3 \, dx \, \tan^2\varphi$$

or, since,

$$J_T = \frac{2}{3} \int\limits_{x_A}^{x_F} y^3 \, dx$$

and,

$$BM_T = \frac{J_T}{\nabla}$$

it follows that

$$M_H = \nabla \, BM_T \, \tan\varphi \quad \text{and, } M_V = \frac{1}{2} \nabla \, BM_T \, \tan^2\varphi \qquad (1)$$

Thus the ship's centre of buoyancy moves parallel to baseline from B_o to B_1 and
away from baseline from B_1 to B_2, where B_o is the upright centre of buoyancy and
B_2 the heeled one, as shown in Fig. 2. Accordingly,

$$M_H = \nabla B_0 B_1, \quad \text{and} \quad M_V = \nabla B_1 B_2 \qquad (2)$$

Fig. 2 Calculation of centre of buoyancy for heeled ship sections

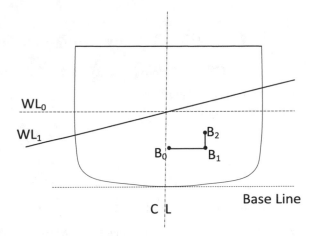

Fig. 3 Explanation of calculation of the position of B

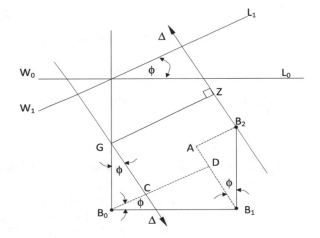

or comparing Eqs. (1) and (2)

$$B_0 B_1 = BM_T \tan \varphi \quad \text{and} \quad B_1 B_2 = 0.5 BM_T \tan^2 \varphi.$$

These two movements can now be translated into a righting lever arm GZ between the heeled lines of action of buoyancy and weight. Geometrically speaking, from Fig. 3:

$$B_0 D = B_0 B_1 \cos \varphi$$

$$A B_2 = B_1 B_2 \sin \varphi$$

$$B_0 C = B_0 G \sin \varphi$$

whilst,

$$GZ = B_0 D - B_0 C + A B_2$$

or,

$$GZ = B_0 B_1 \cos \varphi - B_0 G \sin \varphi + B_1 B_2 \sin \varphi$$

$$GZ = BM_T \sin \varphi - B_0 G \sin \varphi + 0.5 BM_T \tan^2 \varphi \sin \varphi$$

(since $\cos \varphi \tan \varphi = \sin \varphi$).
   Furthermore, since

$$BM_T - B_0 G = GM_T,$$

then

$$GZ = \sin \varphi (GM_T + 0.5 BM_T \tan^2 \varphi). \tag{3}$$

   This formula given in Eq. (3) is the so-called **wall-sided formula**.

Note that if $\tan^2 \varphi$ is small enough the wall-sided formula reduces to $GZ = \sin \varphi \, GM_T$ which is the result obtained earlier for small angles of heel. This actually indicates the angle of heel to which the usual initial stability estimate can be considered valid. A possible limit being, for instance, such that

$$0.5 \, BM_T \, \tan^2 \varphi = 0.02 \, GM_T$$

for which the simpler formula would be 2% in error.

2   Application to Transverse Movement of Weight

If a weight, w, is moved transversely parallel to baseline through a distance d (from A to B) from an initial position in which the ship is upright, the transfer of moment of weight in the heeled state is, as per Eq. (8.2), (see Fig. 4)

$$wd \cos \varphi$$

whilst the righting moment is, as per Eq. (3),

$$\Delta \sin \varphi (GM_T + 0.5 BM_T \tan^2 \varphi)$$

where Δ = total buoyancy inclusive of w and GM_T = effective GM_T with w at the level at which the transfer is taking place (see section on virtual Centre of Gravity). Thus, in the equilibrium state,

$$f(\varphi) = \frac{1}{2} \, BM_T \tan^3 \varphi + GM_T \tan \varphi = \frac{w \, d}{\Delta}.$$

   This is a cubic equation for $\tan \varphi$ which can only be solved by trial-and-error methods or by graph plotting.

Fig. 4 Movement of mass
across ship section

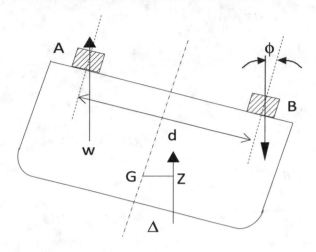

Table 1 Estimation of heeled angle using Eq. (4)

$\tan\varphi$	0.2500	0.2000	0.2200	0.2215
$f(\varphi)$	0.2188	0.1660	0.1863	0.1879
Comment	Too big	Too small		Near enough

Example:

$$w = 1.5\,\text{MN}, \quad \Delta = 120\,\text{MN}, \quad GM_T = 0.75\,\text{m}, \quad BM_T = 4.0\,\text{m} \quad \text{and} \quad d = 15\,\text{m}.$$

This leads to

$$f(\varphi) = 2\tan^3\varphi + 0.75\tan\varphi = 0.1875. \tag{4}$$

Ignoring $\tan^3\varphi$, term gives a first estimate of $\tan\varphi = 0.1875/0.75 = 0.2500$.
   This can be used as a suitable starting value for iteration as shown in the Table 1:
Thus $\varphi = \tan^{-1}0.2215 = 12.49°$.

3   Angles of Loll

A ship for which $GM_T < 0$ is initially unstable and will not float upright. However, there may still be the possibility of the vessel adopting a small angle of heel, either to port or to starboard, for which $GZ = 0$. From the wall-sided formula, Eq. (3),

$$GZ = \sin\varphi \left(-|GM|_T + \frac{1}{2}BM_T\tan^2\varphi\right)$$

and $GZ = 0$ if,

1.

$$\sin \varphi = 0, \quad \text{i.e.} \quad \varphi = 0$$

or

2.

$$-|GM_T| + \frac{1}{2} BM_T \tan^2 \varphi = 0$$

and

$$tan\phi = \pm \sqrt{\frac{2|GM_T|}{BM_T}}$$

The curve of righting levers GZ vs heel angle φ is, in fact, of the form shown in Fig. 5 for a small range of heel angles.

The angles φ_L for which $GZ = 0$ are called **Angles of loll**.

If $\varphi_L = 7°$, for instance $GM_T \geq -0.5\, BM_T \ \tan^2 7° = -0.0075\, BM_T$.

Clearly only small amounts of negative GM_T can be tolerated. It was at one time not at all unusual for vessels loaded with timber deck cargo to go to sea in such a state.

Fig. 5 Angle of loll

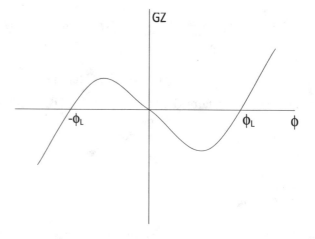

4 Summary

1. The wall-sided formula can be applied to calculate the righting lever GZ at any angle of heel that the wall-sided features, i.e. immersing and emerging parts are triangular wedges of same volume for all sections along the ship, can be assumed.
2. The wall-sided formula is derived and illustrated.
3. Angles of loll are defined.

Large Angle Stability

10

1 The Righting Lever GZ Curve

The hydrostatic stability of a vessel at large angles of heel, possibly up to a complete inversion at $\varphi = 180$, are usually described by a curve of righting levers (GZ) as a function of heel angle φ.

The righting moment at any heel angle is (see Fig. 1)

$$M = \Delta \, GZ.$$

This must be balanced against any moments acting on the ship to heel it over, due to movement of weight, wind loading, ice action, gun fire, etc.

A typical GZ curve has the following characteristics, as shown in Fig. 2:

The curve is of course **anti-symmetric** in that $GZ(-\varphi) = -GZ(\varphi)$. However, normally only one half of the complete curve is drawn.

1. The GZ curve slope at $\varphi = 0$ is related to GM_T since $GZ = GM_T \sin \varphi = GM_T \, \varphi \, (rads)$; i.e. the tangent at $\varphi = 0$ reaches GM_T at 1 radian (57.3°).
2. The GZ curve curves upwards from this tangent as a consequence of the term,

$$\frac{1}{2} \, BM_T \tan^2\varphi \ \sin \varphi$$

   in the wall-sided formula.
3. The curve has a point of inflexion approximately at the angle at which the deck edge immerses.
4. The Range of Stability, may be as low as 20° for a shallow barge, is typically 50°−80° for most mechanically propelled vessels and 130°−160° for sailing vessels (except catamarans). If the craft is completely self-righting, the range must extend to 180°.

© Springer International Publishing AG, part of Springer Nature 2018
P. A. Wilson, *Basic Naval Architecture*,
https://doi.org/10.1007/978-3-319-72805-6_10

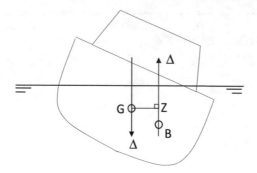

Fig. 1 Righting moments on heeled vessel

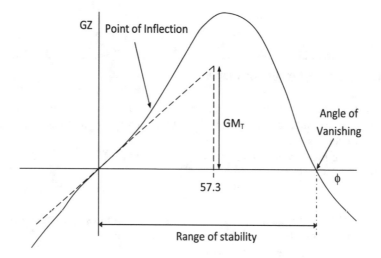

Fig. 2 Righting moment curve

2 Factors Affecting the GZ Curve

2.1 Height of Centre of Gravity

If the height of the Centre of Gravity is raised from G_1 to G_2 (see Fig. 3) this reduces GZ by, exactly by this amount.

$$\delta GZ = G_1 G_2 \sin \varphi \tag{1}$$

Because this change is so simple to compute it is usual practice to calculate GZ for one standard C of G and then to correct for changes of C of G arising in different conditions of loading of the ship using the above formula. Note that the loss of righting lever due to raising C of G, as shown in Fig. 4, is greatest at $\varphi = 90°$ and affects all heel angles to 180°. Clearly the reduction in GM_T reduces the initial slope

Fig. 3 Effects of increasing
G on *GZ*

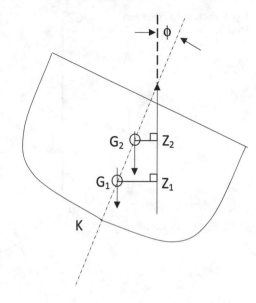

Fig. 4 Effect of position of
G on *GZ* curve

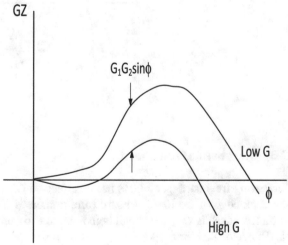

of the curve. If the shift upwards is large enough the vessel becomes unstable upright
and an angle of loll arises (see Fig. 4).

2.2 Increasing Beam

Increasing beam increases J_T since J_T is proportional to B^3 (beam at waterline) and
hence GM_T also. Thus increasing beam increases GZ, both in terms of initial slope
and maximum value. The increase in Range of Stability may not be too dramatic as
freeboard, and superstructure configuration has most effect on this (see Fig. 5).

Fig. 5 Increasing beam and
its effect on *GZ*

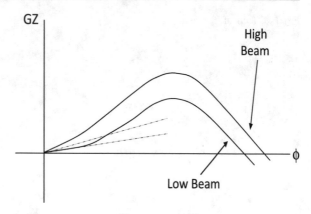

Fig. 6 Increasing freeboard

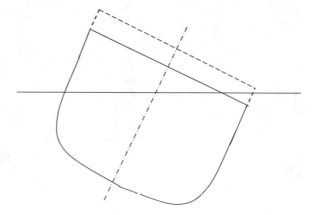

2.3 Increasing Freeboard

Increasing freeboard (see Fig. 6) has no effect on *GZ* at all up to the point at which
the deck edge of the low freeboard form immerses (see Fig. 7). Beyond this point
extra freeboard is very beneficial both in increasing maximum *GZ* and in extending
the Range of Stability.

2.4 Watertight Superstructure

A watertight and **closed** superstructure less than full deck width will offer the same
benefit as extra freeboard once the superstructure begins to immerse (see Fig. 8).
This may well not occur until after the maximum *GZ* for the hull alone. The addition
of a suitable superstructure can have a very beneficial effect on Range of Stability
without making the maximum value of *GZ* too large. This can be useful in providing a
self-righting capability without excessive maximum righting moments that cause the
vessel to self-right too quickly, to the danger of people and gear on board (see Fig. 9).

Fig. 7 Effect of increase of freeboard on *GZ*

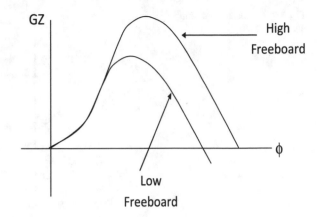

Fig. 8 Heeled ship section

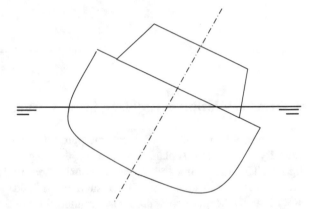

Fig. 9 Stability curve with/without integral superstructure

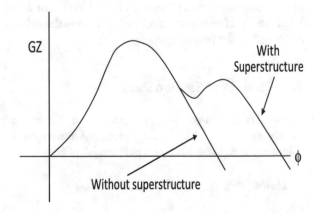

Fig. 10 Digitisation of ship
section

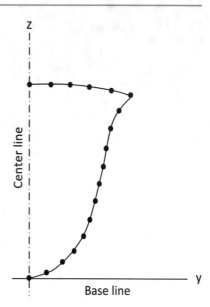

3   The Calculation of Righting Lever Curves

In the past a variety of mechanical integrator devices were used to help perform
stability calculations. A planimeter is a simple integrator to measure areas of closed
figures. The Amsler Integrator simultaneously measured areas and first moments of
area by a similar mechanism. Without such aids a hand calculation would be impos-
sibly lengthy. Modern computers have rendered these devices obsolete. For example the programmes available at the University of Southampton, developed by the
Wolfson Unit for Marine Technology and Industrial Aerodynamics (WUMTIA),
are based on the following method:

3.1   Storage of Section Data

Section Data is stored as (y, z) coordinates digitised randomly round each section on
a body plan (see Fig. 10). Using a suitable numerical integration scheme integrated
values are computed to each input data point for:

1. **Half**-section areas or BONJEAN values

$$a(z) = \int_{o}^{z} y \, dz$$

2. 1st Moment of **half**-section about centreline

$$M_H(z) = \frac{1}{2} \int_{o}^{z} y^2 dz$$

3. 1st Moment of **half**-section about baseline

$$M_V(z) = \int\limits_o^z y\, z\, dz$$

These values are computed and stored in the Hull Definition operation. M_H is referred to as **Horizontal Moment** and M_V as **Vertical Moment**. The integration is an adaptation of Simpson's first Rule suitable for non-equal data spacing.

3.2   Properties of a Full Section at an Angle of Heel

In order to calculate volume and centre of buoyancy, coordinates for the heeled ship section properties are needed for the immersed part of each section.

Using the stored data for each section the heights z_1 and z_2 at which the specified heeled waterline intersects the section are found by interpolation. Once the intersections are located, half-section properties $[y_1, a_1, m_{H_1}, m_{v_1}]$ at z_1 and $[y_2, a_2, m_{H_2}, m_{v_2}]$ at z_2 are also interpolated. In the section sketch (shown in Fig. 11) the left-hand half-section to z_1 leaves out region 1 and this must be added in. On the other hand, the right-hand half-section includes region 2 and this must be subtracted to arrive at immersed section properties.

Treating the moments m_H of the section to the right of centreline as positive and to the left as negative, the overall immersed section properties are:

$$AREA \qquad\qquad A = a_2 + a_1 - \frac{1}{2}\, y_2^2 \tan\varphi + \frac{1}{2}\, y_1^2 \tan\varphi$$

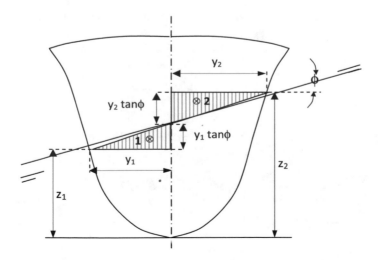

Fig. 11 Calculation of heeled section data from upright Bonjean curves

$$HORIZ\,MOM \qquad M_H = m_{H2} - m_{H1} - \frac{1}{6}\,y_2^3\,\tan\varphi - \frac{1}{6}\,y_1^3\tan\varphi$$

$$VERTICAL\,MOM \quad M_V = m_{V2} + m_{V1} - \frac{1}{2}\,y_2^2\,\tan\varphi\left(z_2 - \frac{1}{3}\,y_2\,\tan\varphi\right) + \frac{1}{2}\,y_1^2\,\tan\varphi\left(z_1 + \frac{1}{3}\,y_1\,\tan\varphi\right)$$

In the moment expressions $(1/3)y_1$ and $(1/3)y_2$ are the horizontal distances of the centroids for wedges 1 and 2, respectively, from centreline. Furthermore $(1/3)y_1\tan\varphi$ and $(1/3)y_2\tan\varphi$ are the vertical distances of the centroids for wedges 1 and 2 above level z_1 and below level z_2, respectively. These properties are required for each section on the body plan.

3.3 Integrated Properties of Immersed Volume

The immersed section properties must now be integrated along the ship length, using Simpson's first Rule, to yield the displacement volume ∇, the distance of the centre of buoyancy $\bar{x}$ (fore and aft from amidships), $\bar{y}$ (transversely from centreline) and $\bar{z}$ (vertically above baseline), as:

$$\nabla = \int\limits_{x_A}^{x_F} A\,dx$$

$$\bar{x} = \frac{1}{\nabla}\int\limits_{x_A}^{x_F} A\,x\,dx\,, \qquad \bar{y} = \frac{1}{\nabla}\int\limits_{x_A}^{x_F} M_H\,dx\,, \qquad \bar{z} = \frac{1}{\nabla}\int\limits_{x_A}^{x_F} M_V\,dx.$$

3.4 The Calculation of GZ

As for the wall-sided formula GZ can now be found geometrically (see Fig. 12) as:

$$GZ = ab + cd$$

$$GZ = Kb + cd - Ka$$

or

$$GZ = \bar{y}\,cos\varphi + \bar{z}\,sin\varphi - KG\,sin\varphi.$$

Hence

$$GZ = \bar{y}\,cos\varphi + (\bar{z} - KG)\,sin\varphi.$$

Fig. 12 Detail of how to calculate the CG for heeled ships

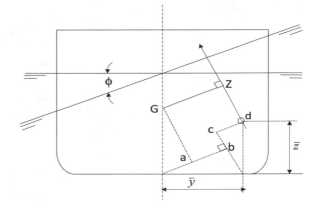

3.5 Varying Draught and Trim

The point at which the heeled waterline cuts for each section along the hull is a function of both mean draught and trim. To cover a complete set of conditions the calculations must be repeated for a range of draughts, a range of trim angles and a range of heel angles. Several hundred calculations may be required—which is why it is not done by hand!

3.6 Cross Curves Calculation Mode

For merchant ships that have a wide range of cargo loading a set (or sets) of **Cross Curves of Stability** are then obtained. For example using the Wolfson Unit Stability programme the quickest way is to run it in the **Fixed Trim** mode over a range of draughts and trims, and to use the information obtained to construct a set (or sets) of **Cross Curves of Stability**.

A set of draughts is chosen for each trim condition, together with a **ROLL CENTRE**. The vessel is then rotated about the roll centre to a series of heel angles (say every 10°), as shown in Fig. 13. Direct calculations of Δ and GZ are carried out for each of the waterplanes defined by this scheme using the algorithm already described.

A set of cross curves of GZ versus Δ can now be drawn in the form shown in Fig. 14. The points on the curves correspond to the chosen waterplanes. The displacement at each waterplane depends on the roll centre. Once the cross curves have been drawn for one choice of KG, values of GZ at each heel angle for that KG and any chosen displacement (Δ_1 say) can be lifted from the curves. The GZ values can be corrected to the appropriate value of KG (KG_1 say) for the loading condition under investigation, as given by Eq. (1). Each loading condition would correspond to different arrangements of cargo in the cargo holds, different amounts of fuel, stores and ballast water, etc.

The programme can also be run in a *free to trim* mode in which Δ, LCG and KG are specified for a particular ship loading condition. For each specified angle of heel

Fig. 13 Effect of heeled ship with respect to different waterlines

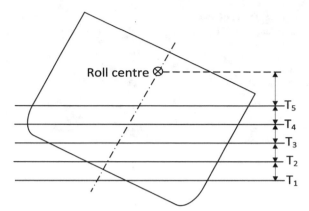

Fig. 14 Cross curves of stability

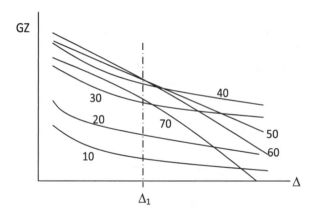

the waterplane is iteratively adjusted, using *TPM* and *MCT* calculations, until the correct displacement and *LCB* are obtained. This usually means carrying out five or so complete calculations at each heel angle in arriving at the right waterplane. This mode of calculation is obviously slower than the fixed trim mode if several loading conditions need investigation. *GZ* curves derived from a set of cross curves at one fixed trim do not differ significantly from free to trim calculations except for vessels with large intact deckhouses or superstructures which cause large changes of trim when immersed at large heel angles.

With large superstructures there may be two equilibrium states with the vessel inverted, as shown in Fig. 15.

4 Dynamical Stability

When a vessel heels over in response to an externally applied heeling moment (for instance due to a wind gust loading) that heeling moment will do work on the vessel. As the vessel rolls against the righting moments generated hydrostatically it will

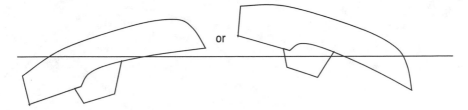

Fig. 15 Inversion of ship with large intact superstructure

Fig. 16 Cylinder rotated by force F

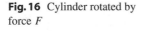

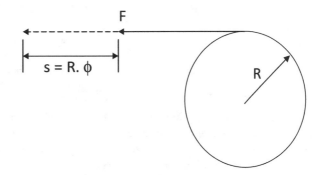

absorb the work done, or a part of it, as a stored or potential energy, to be released again once the hull is allowed to roll back upright. Since heeling and righting moments vary differently with heel angle, with the work done by heeling moments exceeding energy stored in the early stages of heeling, there will also be present a kinetic energy of roll motion related to the rate at which the vessel is rotating in heeling over.

4.1 Basic Concepts

Consider a cylinder being turned by a rope exerting a force F tangentially to the cylinder, as shown in Fig. 16.

If the cylinder rotates through an angle φ radians the rope extends by a distance $s = R\,\varphi$. The work done by whatever is pulling the rope is;

$$W = F\,s = FR\,\varphi.$$

But the moment exerted about the cylinder axis,

$$M = FR;$$

Hence the work can be expressed as:

$$W = M\,\varphi.$$

Fig. 17 Kinetic energy of
rotating particles

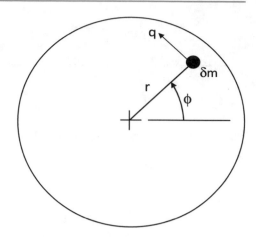

This gives the work done by a moment applied to a rotating body.

An element δm of a rotating body (see Fig. 17), rotating with angular velocity $\dot{\phi}$ rads/sec, at radius r is moving with a speed $q = r\dot{\phi}$ in a peripheral direction.

This element has kinetic energy given by,

$$\delta KE = \frac{1}{2}\,\delta m\;q^2 = \frac{1}{2}r^2\,\delta m\;(\dot{\phi})^2\;.$$

Summing for all elements of the body gives a total kinetic energy due to rotation given by,

$$KE = \frac{1}{2}\left\{\sum r^2\,\delta m\right\}(\dot{\phi})^2 = \frac{1}{2}I\,(\dot{\phi})^2$$

where $I = \sum r^2\,m$ is called the **Moment of Inertia** of the body.

4.2 Application to Ships

In ship terms the work done by an external heeling moment $M(\varphi)$ (not constant, necessarily) is given by:

$$W = \int_{\varphi_1}^{\varphi_2} M(\varphi)\,d\varphi$$

where $\varphi_1 =$ initial heel angle and $\varphi_2 =$ final heel angle.

The potential energy stored against the righting moment $\Delta GZ(\varphi)$ is given by,

$$PE = \Delta \int_{\varphi_1}^{\varphi_2} GZ(\varphi)\,d\varphi$$

Fig. 18 Definition of
dynamical stability

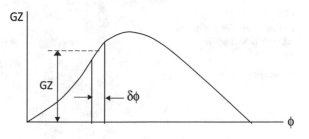

where

$$\int_{\varphi_1}^{\varphi_2} GZ(\varphi)\, d\varphi$$

is the area under GZ curve between φ_1 and φ_2 (measured in metres rads),
As a matter of definition,

$$\int_{0}^{\varphi_1} GZ(\varphi)\, d\varphi$$

is called the **DYNAMICAL STABILITY** of the vessel at angle φ_1 (see Fig. 18).
It is a measure of the ability of the vessel to absorb energy.

If at any stage in the heeling process the work done by an external heeling moment
exceeds the energy stored due to dynamical stability the excess appears as kinetic
energy of roll motion, unless it is dissipated by a damping mechanism, e.g. bilge keels
or stabiliser fins, in addition to the damping originating from the surrounding water.
Note that the work done by a heeling moment can also be expressed by dividing the
heeling moment by the displacement to obtain a **heeling lever**, by analogy to the
righting lever GZ.

Thus,

$$\mathrm{HL}(\phi) = \frac{\mathrm{M}(\phi)}{\Delta} \qquad \text{and} \qquad \mathrm{W} = \Delta \int_{\phi_1}^{\phi_2} \mathrm{HL}(\phi)\, d\phi.$$

4.3   Response to Suddenly Applied Moments

The ship response to an externally applied heeling moment can be analysed on an
energy basis, by comparing a **heeling lever** (HL), related to the applied moment,
with the **righting lever** (GZ) curve, as shown in Fig. 19.

The worst case scenario is when the moment is suddenly applied when the vessel
is at an initial heel angle φ_1 and when the motion is undamped or unresisted. With

Fig. 19 Wind heeling moment and *GZ* variation with ship heel angle

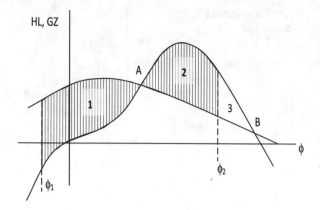

reference to Fig. 19, Points *A* and *B* are possible equilibrium heel angles at which heeling and righting moments balance. Point *A* is a stable position reached in a slowly applied loading or once the kinetic energy of roll has been damped. If the vessel heels beyond *B* it will capsize. The shaded area (1) represents the excess work done over energy stored and, in the unresisted case, represents roll kinetic energy acquired during heeling. Roll velocity is a maximum passing point *A* if the roll is unresisted. The shaded area (2) represents an excess of energy stored over work done during the second phase of the roll motion. As heeling increases beyond *A* the roll kinetic energy is gradually converted to stored energy until at point 2 all the energy is stored and the vessel comes to rest at its maximum heel angle. Subsequently the vessel rolls back towards *A* if $GZ > HL$ at φ_2. It is essential that the roll stops before point *B* is reached. Note that for the unresisted rolling case the shaded areas (1) and (2) are equal and both relate to the roll kinetic energy at point *A*. Area (3), between point *B* and φ_2, represents a margin of safety against capsize, being additional energy storage available prior to reaching *B*.

4.4 Stability Criteria

1. The basic *GZ* curve (see Fig. 2):

   - Acceptable min GM_T.
   - Lowest acceptable maximum GZ and corresponding heel angle.
   - Lowest acceptable Range of Stability.
   - Minimum allowable dynamic stability to, say, 40° heel.

2. Energy Balance Criteria:

   - These consider various suddenly applied moments, for example as in the previous section, and assume unresisted rolling. Moments may be due to wind gust loading, sudden transfer of cargo loads, sudden application of tow rope loads, etc.

- The requirement is in terms of the margin of energy absorption in region (3), which may be required to be say 50% of (1) for adequate safety for certain worst case loading assumptions.
- The down-flooding angle is the smallest heel angle at which any potentially non-watertight opening becomes immersed. Energy balance related margins may be applied to the down-flooding angle rather than the point of capsize.

5 Summary

1. The variation of righting lever GZ with heel angle is illustrated.
2. Influences of various parameters on GZ are discussed. In brief:

   a. High vertical position of C of G has detrimental effects on GZ.
   b. Increasing beam and freeboard and making superstructures watertight have positive effects on GZ.

3. A typical method for calculating GZ variation with heel angle, through the use of cross curves of stability, is illustrated.
4. Dynamic stability is introduced using the energy balance between the work and energy of the heeling and righting moments.
5. Various criteria for the GZ curve are listed.

Flooding Calculations

11

It is important that a ship should be capable of sustaining at least a moderate degree of damage without sinking or capsizing. Ships are usually subdivided internally into watertight compartments to limit the extent of flooding that follows structural damage due to collision, grounding or stress of weather. During the design of the ship calculations are carried out to enable the naval architect to define a satisfactory disposition of major bulkheads to meet suitable safety standards and to examine the consequences of flooding certain spaces, or combinations of spaces, within a ship. With certain classes of ship, principally those carrying passengers, there are legal requirements to meet certain standards of subdivision. For other ships there is no legal necessity to meet such requirements, but the ships are allowed to load to a deeper draught if they do. Naval vessels must obviously meet very high standards of subdivision in order to fulfil their military role. Leisure craft are a special case in that foam filled buoyancy spaces are frequently used to ensure the craft survives damage. Nevertheless calculations are still needed to obtain an adequate disposition of such *buoyancy*.

1 Definitions Used in Subdivision

1.1 Bulkhead Deck

The ship is divided into watertight compartments up to a deck designated as the BULKHEAD DECK. Above this level the vessel is treated as being freely floodable - as a *worst case* approach to the problem. Openings in watertight bulkheads must be closed by doors capable of sustaining water loads on one side. For safety these doors must be normally closed at sea. Thus, access between compartments is very restricted below the bulkhead deck. Different decks may be designated as the bulkhead deck in different parts of the ship.

© Springer International Publishing AG, part of Springer Nature 2018
P. A. Wilson, *Basic Naval Architecture*,
https://doi.org/10.1007/978-3-319-72805-6_11

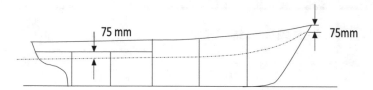

Fig. 1 Load waterline

1.2 Margin Line

The margin line is a fair curve drawn 75 mm below the bulkhead deck at side of ship. Where there is a change of level of bulkhead deck the line is drawn in a form as shown in Fig. 1.

1.3 Compartment Permeability (μ)

No compartment is ever completely empty. Only a certain fraction is available for flooding.

$$\mu = \frac{\text{Volume available for flooding}}{\text{Total compartment volume}} \quad (1)$$

Whilst major impermeable items in a compartment may be identified individually, calculations are simplified by considering the remainder of the space to be *uniformly permeable* (i.e. every part of the space has the same permeability).

Typical values of μ are:

- Cargo Spaces μ = 0.63
- Machinery Spaces μ = 0.85
- Passenger Spaces μ = 0.98.

1.4 Floodable Length

At any point along the length of the ship it is possible to define a maximum length of compartment which, if flooded to the appropriate permeability, will result in the vessel floating at a waterline which touches, but does not immerse, the margin line. This maximum compartment length is called the FLOODABLE LENGTH (see Fig. 2).

The relationship between **permitted** compartment lengths and **floodable** lengths are dealt with when legal requirements are discussed.

Fig. 2 Floodable length

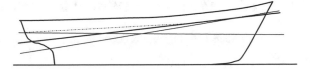

2   Added Weight and Lost Buoyancy Calculation Methods

There are two basic methods of calculation applied to flooding problems, depending as to how the **flood water** is treated in the calculations. Each needs to be applied in a logically consistent way:

2.1   Added Weight Method

1. The vessel is treated as *intact* for the purpose of calculating buoyancy, centres of buoyancy and the position of the transverse metacentre.
2. The flood water is treated as an *additional weight* on board which alters the displacement weight (or mass) and the position of the Centre of Gravity, both longitudinally and vertically.

2.2   Lost Buoyancy Method

1. The vessel is treated as though the flooded compartment is *no longer part* of the ship, the buoyancy provided within the flooded compartment to the original undamaged waterline being considered *lost* to the vessel.
2. The weight and Centre of Gravity of the vessel is taken to correspond to the *undamaged* state and to remain *unchanged* after flooding.
3. Hydrostatic particulars for the vessel are adjusted to correspond to the **residual** ship values. To calculate sinkage and trim *modified* values of ∇, LCB, LCF, TPM, MCT etc. are required. The vessel will change draught and trim in such a way that the **residual ship** has the same ∇, LCB as the intact vessel.

The **residual ship** is, of course, the intact ship **less the permeable** part of the flooded space. Where $\mu < 1.0$ the flooded space will make a partial contribution to both volume and waterplane properties.

The **added weight** method is the most direct route for calculations of flooding to a prescribed waterline (e.g. a line touching the margin line). It is thus used in calculations of *floodable length*.

The **lost buoyancy** method is the most direct route for finding flooded draughts and trims consequent upon flooding specific spaces within the ship.

3 Flooding to a Specified Waterline

Using the added weight route proceed as follows:

- Calculate the local draught at each section along the hull corresponding to the chosen waterline. Find the corresponding immersed area of each section (see Fig. 3).
- Compute the flooded displacement volume ∇_1 and the longitudinal centre of buoyancy LCB_1 for the whole vessel to the chosen waterline, shown in Fig. 3 (by integrating the sectional areas and moments about, say, amidships).
- Find the flood water volume and Centre of Gravity by comparison with the corresponding intact ship values (∇_0 and LCB_0):-
Flood Water volume

$$V_F = \nabla_1 - \nabla_0 \tag{2}$$

Compartment volume to flooded WL,

$$V_c = \frac{\nabla_1 - \nabla_0}{\mu} \tag{3}$$

Longitudinal Centroid of Flood Water,

$$\lg = \frac{\nabla_1 \ LCB_1 - \nabla_0 \ LCB_0}{V_F}. \tag{4}$$

The remaining problem is to locate the end bulkheads of the compartment to enclose the required flood water volume and having the right centroid. The distance between these bulkhead positions defines the floodable length for that part of the ship (provided the chosen waterline touches the margin line).

Trial positions for the required after bulkhead (A) and forward bulkhead (F) of a compartment of length l and centred at x from amidships can be superimposed on a curve of immersed sectional areas $a(x)$ to the chosen flooded waterline. These trial positions are taken half length either side of compartment centroid. As a first approximation choose the flood water centroid as the compartment centre and base compartment length l on the sectional area at this point, i.e. $l = \frac{V_c}{a(x=lg)}$. Ordinates of areas at equal spacing along the compartment length can be measured from the curve and integrated to obtain the compartment volume to the flooded waterline, together

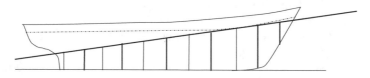

Fig. 3 Immersed areas

Fig. 4 Area curve

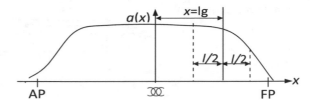

Fig. 5 Floodable length curve

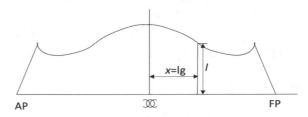

with the corresponding centroid of volume. The volume and centroid so obtained is compared to V_c and lg. Values of x and l are altered, and the calculation is repeated until the two sets of figures agree sufficiently closely (Fig. 4).

3.1 Constructing a Floodable Length Curve

By carrying out the above calculations for a large number of different flooded water-lines, each touching the margin line, a large number of pairs of values of (lg, l) can be obtained. For each flooded waterplane there will be a particular compartment centre $(x = lg)$ and compartment length (l) defining a possible compartment which, if flooded, would bring the ship down to that waterplane. The values are, of course, specific to a particular choice of permeability. A curve of l versus lg is called a **floodable length curve**, shown in Fig. 5. Floodable length curves for a suitable range of permeabilities are used in the process of deciding where watertight bulkheads should be positioned in the ship.

4 Flooding a Specified Compartment

To obtain the final flooded draught and trim resulting from the flooding of a com-partment between specified bulkheads it is best to use the **lost buoyancy** method of calculation. It is normal to assume that flooding is symmetric about centreline, so that heeling does not arise, and to consider only the final stage of flooding for which water levels inside and outside the ship are equal. The problem will be treated as a first order sinkage and trim problem.

W_0L_0 is the intact waterline and W_1L_1 is the flooded waterline. The lost buoy-ancy method treats the floodable part of the compartment as being outside the ship.

Fig. 6 Flooded compartment

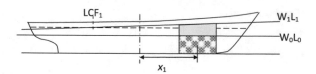

Consider first the deficit of displacement volume and moment of volume at the intact waterplane W_0L_0 (see Fig. 6):

Compartment Volume below $W_0L_0 = V_1$
Flood water in compartment below $W_0L_0 = \mu \times V_1$
                                                     = Deficit of displace-

ment volume

                                                     $= \delta\nabla$

This deficit will have its centroid at x_1 (say) from amidships.

In order to recover the lost volume and lost moment of volume about amidships the vessel will undergo a **parallel sinkage** together with a **change of trim**. There will be a certain amount of displacement volume gained in the flooded compartment as follows:

Compartment Volume between W_0L_0 and W_1L_1 $= V_2$
Flood water between W_0L_0 and W_1L_1 $= \mu \times V_2$
Gain in displacement volume between W_0L_0 and W_1L_1 $= (1 - \mu)V_2$

This gain of volume can be correctly accounted for by using residual waterplane properties in computing TPM, LCF, MCT etc. for the damaged ship. The required residual waterplane properties are:

Residual Waterplane Area $A_1 = A_0 - \mu \times a$
Residual 1st Moment of WP Area about amidships $A_1 \times LCF_1 = A_0 \times LCF_0 - \mu \times m$

Residual 2nd Moment of WP Area about amidships $J_{L1} = J_{L0} - \mu j_L$
where, a = Compartment waterplane area,
          m = Compartment 1st Moment of area about amidships,
          j_L = Compartment 2nd Moment of area about amidships.
A_0, LCF_0, J_{L0} = intact ship waterplane properties,
A_1, LCF_1, J_{L1} = residual ship waterplane properties.

In order to compute a residual MCT, J_{L1} must be corrected to give J_{L1} residual 2nd. Moment about LCF_1 using the parallel axis theorem.

As usual in sinkage and trim problems the lost volume $\delta\nabla$ is recovered by a parallel sinkage:

$$\delta T = \frac{\delta\nabla}{A_1} = \mu\,\frac{V_1}{A_1} \tag{5}$$

The additional $\delta\nabla$ is added at LCF_1, thus introducing a trimming moment which is removed by a change of trim about LCF_1 (see Fig. 7). Once the sinkage and trim have been found, final draughts forward and aft follow in the usual way.

Fig. 7 Change of
displacement at LCF

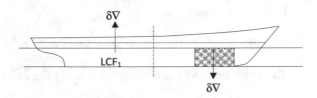

5   Corrections for Sinkage and Trim

If the sinkage and trim produced by flooding is too large to use first-order methods
based on the properties of the waterplane W_0L_0 then it becomes necessary to use an
iterative procedure to find the final draughts on flooding a particular compartment.

1. Use first-order lost buoyancy calculations as above to determine approximate
   flooded draughts.
2. Use added weight calculations to find ∇, LCB to approximate flooded WL. from
   stage 1.
3. By comparing ∇, LCB to values inclusive of flood water in compartment estimate
   an error in ∇ to trial WL.
4. Correct estimated flooded draughts using residual WP properties for trial WL and
   repeat from stage 2 as needed.

This procedure is only suitable for computer-based calculations.

6   Example: Added Weight Calculation

A rectangular box-shaped vessel had dimensions $LBP = 140$ m, $B = 20$ m and floats
at a level keel draught $T = 8.0$ m when intact.

Find the required end bulkhead positions of a compartment of uniform permeability $\mu = 0.70$ which, when flooded, results in a forward draught of 13.0 m and an
after draught of 6.0 m.

For the intact vessel,

$$\nabla_0 = 140 \times 20 \times 8 = 22400 \, \text{m}^3 \tag{6}$$

For the flooded vessel the mean draught is

$$\frac{13.0 + 6.0}{2} = 9.5 \, \text{m}, \tag{7}$$

LCF is at midships and the displacement is

$$\nabla_1 = 140 \times 20 \times 9.5 = 26600 \, \text{m}^3. \tag{8}$$

Hence, flood water volume is

$$V_F = \nabla_1 - \nabla_0 = 4200\,\text{m}^3. \tag{9}$$

Therefore, required compartment volume to flooded WL is

$$V_c = \frac{V_F}{\mu} = \frac{4200}{0.70} = 6000\,\text{m}^3. \tag{10}$$

For a box-shaped vessel,

$$J_L = \frac{L^3 B}{12}, \ BM_L = \frac{J_L}{\nabla} = \frac{L^2}{12T}. \tag{11}$$

For the flooded vessel,

$$BM_L = \frac{140^2}{12 \times 9.5} = 171.93\,\text{m}. \tag{12}$$

At a trim of $13.0 - 6.0 = 7.0\,\text{m}$ by bow, the change of LCB (which is also LCB distance from amidships, as this is a box) can be calculated as

$$LCB = BM_L \times \frac{Trim}{L_{BP}} = 171.93 \frac{7.0}{140} = 8.596\,\text{m fwd}. \tag{13}$$

Please note as the flooded displacement for this rectangular box is of trapezoidal form LCB can also be calculated from the trapezoid centroid.

The volume trimming moment is $26600 \times 8.596 = 228670\,\text{m}^4$
= moment due to flood water.

Hence, LCG of flood water is

$$\frac{228670}{4200} = 54.44\,\text{m fwd} \tag{14}$$

Which is equal to $lg = $ Centroid of compartment volume below flooded WL.
At the flood water LCG the local draught is

$$9.5 + \frac{7.0}{140} \times 54.44 = 12.22\,\text{m}; \tag{15}$$

Thus, the immersed cross-sectional area is $20 \times 12.22 = 244.44\,\text{m}^2$ at the flood water LCG. Taking this area as an approximate mean area for the flooded space gives a first approximate compartment length as,

$$l = \frac{6000}{244.4} = 24.55\,\text{m}. \tag{16}$$

Fig. 8 Trapezium

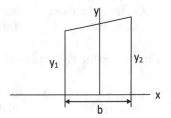

Taking the flood water LCG as approximately the mid-length of the flooded compartment gives a first estimate of the end bulk heads at,

$$54.44 \pm \frac{24.55}{2} = 42.17 \text{ m fwd and } 66.71 \text{ m fwd.} \tag{17}$$

Strictly speaking, because of the trim of the vessel, the flood water LCG is actually slightly forward of the compartment mid-length and both bulkheads are slightly too far forward. In this example the error is approximately 0.20 m:

For a trapezium the centroid is at (see Fig. 8),

$$\frac{b}{6}\left(\frac{y_2 - y_1}{y_2 + y_1}\right). \tag{18}$$

Thus the distance of the compartment CG from the mid-length can be found on substituting for compartment length and draughts at the bulkheads. Thus the bulkheads are more accurately at 42.0 and 66.5 m.

7 Example: Lost Buoyancy Method

A rectangular box-shaped vessel has dimensions $LBP = 140$ m, $B = 20$ m and floats at a level keel draught $T = 8.0$ m when intact.

Find the forward and aft draughts in the final flooded state if a full width compartment with end bulkheads at 40 and 65 m forward of amidships is open to the sea above the double bottom tank space. The double bottom tank is 1.5 m deep. The compartment is 0.70 permeable.

The floodable compartment depth to the intact waterplane is $8.0 - 1.5 = 6.5$ m.

Thus lost displacement volume is $\delta \nabla = 20 \times (65 - 40) \times 6.5 \times 0.70 = 2275 \text{ m}^3$.

The centroid of the lost volume is at,

$$\frac{65 + 40}{2} = 52.5 \text{ m fwd of amidships.} \tag{19}$$

The various waterplane properties required are as follows:

Intact Waterplane

- Area: $A_0 = 20 \times 140 = 2800 \text{ m}^2$

- Centroid: $LCF_0 = 0$ m amidships
- 2nd Momt: $J_{L0} = \frac{1}{12} \times 140^3 \times 20 = 4.5733 \times 10^6$ m^4

Compartment Waterplane

- Area: $a = 20 \times (65 - 40) = 500$ m^2.
- Centroid: $lcf = 52.5$ m fwd of amidships.
- 2nd Momt: $j_L = \frac{1}{12} \times 25^3 \times 20 + 500 \times 52.5^2 = 1.4042 \times 10^6$ m^4 about amidships.

Residual Waterplane

- Area $A_1 = 2800 - 0.70 \times 500 = 2450$ m^2
- 1st Mom about midships, $A_1 \times LCF_1 = 0 - 0.70 \times 500 \times 52.5 = -18375$ m^3(aft)

Thus $LCF_1 = \frac{-18375}{2450} = 7.5$ m abaft amidships.

- 2nd Mom about midships $J_{L1} = 4.5733 \times 10^6 - 0.70 \times 1.4042 \times 10^6 = 3.5904 \times 10^6$ m^4
- 2nd Mom about LCF_1 $J_{L1} = 3.5904 \times 10^6 - 2450 \times 7.5^2 = 3.4525 \times 10^6$ m^4.

Assuming $GM_L \approx BM_L$ then,

$$MCT = \nabla \frac{BM_L}{L_{BP}} \text{ m}^4/\text{m (in volume units)} \tag{20}$$

Thus,

$$MCT = \frac{J_L}{L_{BP}} = \frac{3.4525 \times 10^6}{140} \text{ m}^4/\text{m} = 24661 \text{ m}^4/\text{m}. \tag{21}$$

We are now ready to calculate the sinkage and trim required to restore the lost buoyancy: Parallel sinkage $\delta T = \frac{\delta \nabla}{Area} = \frac{2275}{2450} = 0.929$ m.
Distance of flood water LCG ahead of $LCF_1 = 52.5 + 7.5 = 60.0$ m.
Hence, volume moment to change trim is

$$= 2275 \times 60 = 136500 \text{ m}^4, \tag{22}$$

and bow down trim change =

$$\frac{136500}{24661} = 5.535 \text{ m}. \tag{23}$$

Hence the flooded draughts are:

$$T_F = 8.0 + 0.929 + 5.535 \times \frac{(70 + 7.5)}{140} = 12.00 \text{ m} \tag{24}$$

$$T_A = 8.0 + 0.929 - 5.535 \times \frac{(70 - 7.5)}{140} = 6.46\,\text{m}. \tag{25}$$

8 Summary

- The basic definitions used in subdivision, i.e. bulkhead deck, margin line, permeability and floodable length are made.
- The differences between added weight and lost buoyancy calculation methods for flooding problems are indicated and illustrated with examples.
- Added weight method is used to obtain the amount of flood water, and its centroid when the final waterline is known, i.e. flooding to a specified waterplane. This method is, therefore, used to obtain floodable length curve.
- Lost buoyancy method is used to obtain the final waterline following flooding when the amount and centroid of floodwater is known - i.e. flooding a specified compartment. The most important aspect of the lost buoyancy method is the residual ship and the evaluation of residual water plane properties, used to define *TPM* or *TPC* and *MCT* for the residual ship.
- Guidelines are given for calculations involving large changes of draught and trim, using iterative solution.

End on Launching and Launching Calculations

12

In order to ensure a safe launch, calculations are carried out to confirm that:-

1. The vessel will *not tip bow up about the after end of the ways*. This requires that, once G has passed the way end, the moment of buoyancy about the way end always exceeds the moment of weight about the way end (see Figs. 1 and 7).
2. The *maximum loads on the fore poppet* (cradle) are sustainable both by the poppet structure and by the hull. If necessary internal shoring/temporary structure may be used to strengthen the hull locally. Max poppet load occurs at the point of *stern lift* where the moment of buoyancy about the fore poppet has risen to equal the moment of weight about the fore poppet (see Figs. 6 and 7).
3. The *fore poppet load drops to zero before the fore poppet passes the way end*. This is to ensure that the bow docs not forcibly drop off thc way end, possibly resulting in the bow pitching down onto the river bed with consequent structural damage (see Fig. 7). The fore poppet load $F = W - B$.
4. The vessel must remain statically stable, despite the load on the fore poppet, for the whole travel after the point at which the stern lifts. The least stable position is at the point of stern lift.
5. The pressures between the sliding and ground ways should be sustainable by the lubricating grease (see Fig. 2). This implies limitations on both the average pressure over the whole of the contact area and on the peak pressure (occurring at one end of the contact area).
6. The speed achieved during the launch must be controlled and drag chains must be chosen to stop the vessel within the limits of safe travel defined usually by the width of the river.

© Springer International Publishing AG, part of Springer Nature 2018
P. A. Wilson, *Basic Naval Architecture*,
https://doi.org/10.1007/978-3-319-72805-6_12

Fig. 1 Draught
measurements

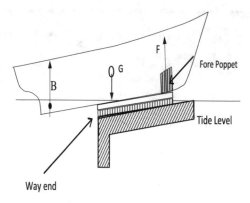

Fig. 2 Draught
measurements

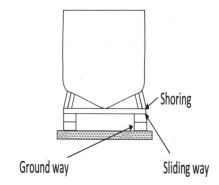

---

1   Ground Way Geometry

1.1   Straight Ways

α = keel declivity
β = way declivity

Typically α = 1:20 and β is a permanent feature of slipway (see Fig. 3). α is usually less than β and can be chosen by setting up the building blocks prior to commencement of building. It has implications at launch, so needs quite early thought!

1.2   Cambered Ways

Camber should be circular to ensure proper mating of sliding and ground ways over full length of travel (see Fig. 4). Ship changes trim (α changes) as it travels down ways. Since camber is small, it is sufficient to assume parabolic camber for purposes of launch calculations.

Fig. 3 Straight ways

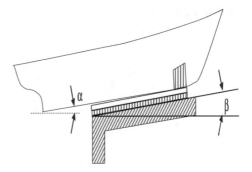

Fig. 4 Cambered ways

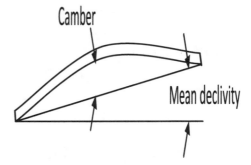

---

2   Launching Calculations

Launching curves are plotted to a base of travel along the ways for a suitable range of tide heights covering the expected launch period. At each point on the curves draughts are established at each body plan section and, as a minimum, **Bonjean** data obtained. It can be also desirable to obtain waterplane offsets and vertical moments in order to compute stability information. Calculations fall into two parts: (a) Prior to stern lift. (b) Post stern lift.

2.1   Prior to Stern Lift

Total load on ways $= W - B$. Assume a linear pressure variation along, ways in contact with ship, chosen to balance buoyancy and way force loads about $C.G$. This is used to calculate mean way pressure and identify maximum way pressure (either at fore or after end of ways, depending on travel), as shown in Fig. 5.

Fig. 5 Straight ways

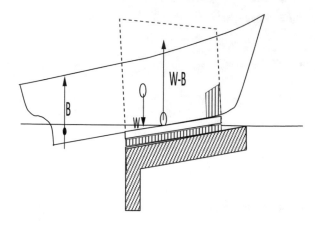

Fig. 6 Prior to Stern lift

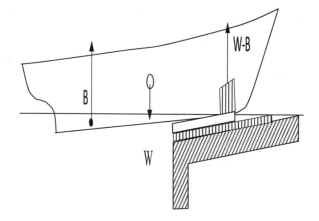

2.2 Post Stern Lift

Vessel trims about fore poppet until moment of buoyancy about poppet balances moment of weight (see Figs. 6 and 7). Calculations assume a range of trim values and cross plot moment curves to identify balance point.

2.3 Launch Curves

Calculations assume static equilibrium throughout launch process. The relevant curves are sketched in Fig. 7.

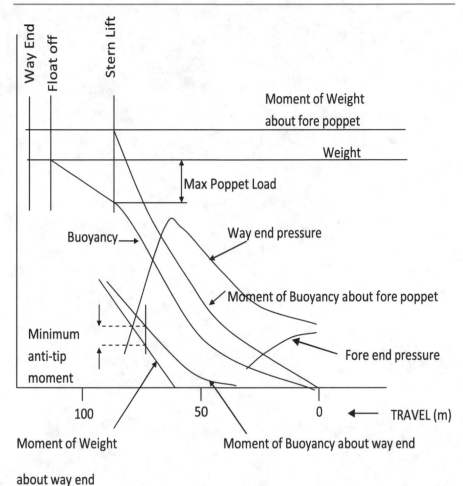

Fig. 7 Launching curves

3 Summary

- The fundamental aspects of end launching are summarised.
- The basics of way geometries are discussed.
- The end launching calculations are described and graphically illustrated. Particular attention is drawn to:

  - ensuring the vessel does not tip bow up about the after end of the ways
  - maximum fore poppet load, occurring at point of stern lift
  - vessel floats off (i.e. fore poppet load becomes zero) before the fore poppet passes the way end.

Stability Assessment Methods (Deterministic and Probabilistic)

13

1 Background

1.1 IMO

One overarching requirement has been the way to improve safety at sea by developing international binding regulations that are followed and adhered to by all nations that operate ships. From the mid-nineteenth century, for example the 1854 Merchant Shipping Act, a number of treaties were developed and adopted, usually following a major shipping disaster. This initially led to the formation of the SOLAS (Safety Of Life At Sea) Convention following the sinking of the *Titanic* in 1912 on its maiden voyage from Great Britain to the USA. Following this disaster and further developments the call from several countries and following the establishment of the United Nations was that in an international conference in Geneva in 1948, the adoption of the Inter-Governmental Maritime Consultative Organisation (IMCO), which eventually changed its name to International Maritime Organisation (IMO) in 1982. The IMCO convention came into force in 1958, and the organisation first met in 1959. The IMO headquarters are in central London.

1.2 Ship Stability Developments

Recent developments in the design of passenger ships have led to larger and larger vessels with an increasing capacity for carrying more passengers and crew. Modern cruise liners have been designed to carry several thousand people, and even though accidents involving such large passenger ships are rare, if a serious accident should occur, its consequences could be disastrous. The safety of large passenger ships is thus an increasingly important issue. Some previous catastrophes involving a large number of fatalities on passenger ships are the collision of the *Titanic* in 1912 (more

© Springer International Publishing AG, part of Springer Nature 2018
P. A. Wilson, *Basic Naval Architecture*,
https://doi.org/10.1007/978-3-319-72805-6_13

than 1500 fatalities), the collision between *Andrea Doria* and *Stockholm* in 1956 (combined fatalities of 51), the collision of the *Admiral Nakhimov (previously Berlin III)* with *Pyotr Vasev* in 1986 (425 fatalities), the capsizing of the *Herald of Free Enterprise* in 1987 (193 fatalities), the collision and subsequent fire and sinking of the *Dona Paz* in 1987 (4386 fatalities), the fire on the *Scandinavia Star* in 1990 (158 fatalities), the foundering of the *Estonia* in 1994 (852 fatalities), the fire and subsequent sinking of the *Dashun* in 1999 (282 fatalities) and more recently the sinking of *Sewol* in 2014 (476 fatalities). These are just some examples of major accidents involving passenger ships, and although all the accidents are characterised by a set of very particular circumstances that lead to the catastrophe, they serve as good examples of the grave consequences that might result from passenger ship accidents.

The SOLAS Convention in its successive forms is generally regarded as the most important of all international treaties concerning the safety of merchant ships. The first version was adopted in 1914, in response to the *Titanic* disaster, the second in 1929, the third in 1948, the fourth in 1960 and the fifth in 1974. The main objective of the SOLAS Convention is to specify minimum standards for the construction, equipment and operation of ships, compatible with their safety. Flag states are responsible for ensuring that ships under their flag comply with its requirements, and a number of certificates are prescribed in the convention as proof that this has been completed. Control provisions also allow Contracting Governments to inspect ships of other Contracting States if there are clear grounds for believing that the ship and its equipment do not substantially comply with the requirements of the convention—this procedure is known as Port State Control. The current SOLAS Convention includes Articles setting out general obligations, amendment procedure and so on, and followed by an Annex, it is divided into the following chapters:

1. Chapter I - General Provisions,
2. Chapter II,

   a. Chapter II-1 - Construction - Subdivision and stability, machinery and electrical installations,
   b. Chapter II-2 - Fire protection, fire detection and fire extinction,

3. Chapter III - Life-saving appliances and arrangement,
4. Chapter IV - Radiocommunications,
5. Chapter V - Safety of navigation,
6. Chapter VI - Carriage of cargoes,
7. Chapter VII - Carriage of dangerous goods,
8. Chapter VIII - Nuclear ships,
9. Chapter IX - Management for the safe operation of ships,
10. Chapter X - Safety measures for high-speed craft,
11. Chapter XI,

  a. Chapter XI-1 - Special measures to enhance maritime safety,
  b. Chapter XI-2 - Special measures to enhance maritime security,

12. Chapter XII - Additional safety measures for bulk carriers,
13. Chapter XIII - Verification of compliance,
14. Chapter XIV - Safety measures for ships operating in polar water.

The SOLAS 1929 and up to SOLAS 1960 Convention set down the series of requirements for the number and arrangements of watertight bulkheads and ships' stability after damage. It was recognised that the semi-empirical nature of the deterministic approaches required and worked from characteristics of previously known disasters. In SOLAS 1960 Wendel talked about the fundamental idea of a probabilistic method of assessment of the watertight subdivision enabling different scenarios of possible damage situations to be investigated. This was adopted some years later in SOLAS 1974 in Resolution *A*265.

Since the SOLAS 1974 Convention a number of consolidation editions have been approved by IMO. For example one that is of direct interest of this chapter is SOLAS 2009. References [1–4]. In order to provide an easy reference to all SOLAS requirements applicable from 1 July 2009, this edition presents a consolidated text of the SOLAS Convention, its Protocols of 1978 and 1988 and all amendments in effect from that date.

The fully updated 2009 edition features a number of new SOLAS regulations, adopted after the last consolidated edition of the convention was published. The SOLAS provisions for corrosion protection have been updated and expanded, and the new requirements are incorporated in Chap. II-1. Furthermore, Chap. II-1 was comprehensively revised to include probabilistic requirements for subdivision and damage stability and now also has a new Part F concerning alternative designs and arrangements. The annex to the convention regarding the SOLAS forms of certificates contains the fully revised safety certificates for nuclear, passenger and cargo ships, and the list of certificates and documents required to be carried on board ships, as revised, is also added.

This publication, compiled by the Secretariat to provide an easy reference to SOLAS requirements, contains a consolidated text of the 1974 SOLAS Convention, the 1988 SOLAS Protocol and all subsequent amendments thereto in force as on 1 July 2009. SOLAS Amendments 2006 includes additional amendments in the form of Resolution MSC.201(81) and Annex 3 of Resolution MSC.216(82) (pp. 1–6 and 88–101, respectively), which will enter into force on 1 July 2010. As these amendments are not in SOLAS 2009, for easy reference they have been added here to the publication.

There are more recent consolidated editions of SOLAS the latest being SOLAS 2016, but since we are concerned in this textbook with stability issues, SOLAS 2009 is the one that is often quoted as most relevant.

1.3   History of the Development of the Probabilistic Methodology

Two major accidents of Ro-Ro Passenger ships in Europe, namely the tragic losses of *Herald of Free Enterprise* in 1987 and *Estonia* in 1994, put the work of IMO on the harmonisation of existing damage stability rules for some time on hold. As a matter of urgency, IMO's SLF (stability, load lines and fishing vessels) and MSC (Maritime Safety Committee) committees addressed first a revision of the existing deterministic 1990 and 1992 amendments of SOLAS damage stability regulations of passenger ships to account for the water on deck problem of Ro-Ro passenger ships. After the adoption of the particular enhanced deterministic requirements in the SOLAS Convention of 1995, IMO relevant committees brought back the harmonisation of damage stability rules on the regulatory agenda, and a first proposal for a revision of SOLAS Chap. II-1 Parts A, B and B-1 was discussed at the Intersessional IMO-SLF42 meeting in 1998.

During this research period to determine the future direction, a team of European industries, classification societies, universities and research establishments, administrations and others proposed to the European Commission and received funding for the research project, HARDER (2000 → 2003). This project's main objective was to generate knowledge in the general field of ships' damage stability by systematic fundamental and applied research and to clarify a variety of technical issues of great importance to the harmonisation work of the tasked IMO-SLF subcommittee. During the HARDER project, the new harmonised damage stability, probabilistic concept, known as the SLF42 proposal, under development at IMO, was systematically evaluated and an improved proposal was introduced for discussion at IMO, known as the HARDER-SLF46 proposal. Both concepts were extensively discussed in IMO-SLF 46/INF.5. In September 2003, the work of harmonisation was actually completed and some late concerns were raised at IMO, particularly in relation to the apparent severe impact of the proposed new harmonised damage stability regulations on the design (and economy) of very large passenger ships, bringing the formal approval of relevant regulations again to a hold. The IMO-MSC instructed the SLF subcommittee to reconsider the issues of concern, and as a result, a series of new studies were carried out addressing particularly the damage stability of large passenger ships. Related proposals for amendments were submitted for consideration to IMO-SLF46 and MSC78.

On the basis of the work of the International Correspondence Group of IMO-SLF46/47, the HARDER-SLF46 proposal was revisited and led to the SLF47 proposal that was essentially approved in September 2004 as IMO-SLF47 and shortly after at IMO-MSC79. This proposal was, however, once more revised with respect to the large ships assessment method on the way from the MSC79 to the MSC80 meeting in May 2005, where it was finally adopted. It is noted that the finally adopted MSC80 probabilistic damage stability assessment concept was to apply to all new dry cargo and passenger ships constructed after 1 January 2009.

2 Damage Stability Calculations

Calculations of stability of damaged ships are complicated and tedious. At present, two different analysis concepts are applied: the *deterministic* concept and the *probabilistic* concept. For both concepts, the damage stability calculations normally are made according to the method of lost volume or lost buoyancy. Unfortunately, the collision resistance is not considered when assessing damage stability and vessels with strengthened side structures are treated in the same way as single-hulled ships. A deterministic approach is a perfectly predictable approach. That is, the approach follows a completely known rule, e.g. a fixed procedure, so that a given input will always give the same output. The states of a system described by a deterministic approach may be numbers specifying physical characteristics of the system, for instance observables such as length or mass. The current damage stability regulations are summarised in Table 1.

For deterministic damage stability of ships (DS), these regulations are said to follow a deterministic approach because the ship must survive a predetermined amount of damage [5]. Deterministic damage stability is all about ensuring that a ship is *safe enough*. It has been the dominating method for a long time, and the DS regulations are still in use today for various reasons. First of all, the probabilistic damage stability (PS) regulations do not cover all ship types. Secondly, despite the DS regulations well-known conservatism, it has a long track-record, and society in general tends to trust well-proven methods. A, C and D in Table 2 as in previous chapters of this textbook follow a deterministic approach. As mentioned, these three regulations apply to different types of passenger ships. In addition to the SOLAS-74 standard,

Table 1 Current damage stability regulations

A	Passenger vessels pre-2009	Deterministic
B	The Stockholm agreement	Probabilistic
C	Safe return to port regulations	Deterministic
D	Damaged stability requirements - Type A/B vessels (pre-2009)	Deterministic
E	SOLAS damaged stability rules post-2009	Probabilistic

Table 2 IMO instruments containing deterministic damage stability

	Regulatory framework	Application area
A	ICCL-66	Cargo ships and tankers with reduced freeboard
B	MARPOL-73/78	Tankers carrying cargo oil
C	IBC Code	Ships carrying dangerous chemicals in bulk
D	IGC Code	Ships carrying liquefied gases in bulk
E	HSC Code	High-speed craft

which regulates ordinary passenger ships, Table 2 provides an overview of other IMO instruments that contain DS provisions [6].

This method is based on damage assumptions such as damage to length, to transverse extent and to vertical extent. Depending on the ship type or potential risk to the environment resulting from the type of cargo carried, compliance with a required compartment status must be proved. The deterministic concept applies to chemical and liquefied gas tankers, bulk carriers, offshore supply vessels, high-speed craft and special-purpose ships.

2.1 Damage Extent

The damage extent comprises the longitudinal-, transverse- and vertical extent of the damage. The longitudinal damage extent is determined by the length of the ship, L, as calculated by Eq. 1 (Patterson and Ridley [7]):

$$\text{Longitudinal damage extent} = min(3 + 0.03 \times L, 11)\,[\text{m}] \tag{1}$$

The definition of the ship length goes all the way back to the International Convention on Load Lines, 1966 (ICCL-66), which states: *Length means 96% of the total length on a waterline at 85% of the least moulded depth measured from the top of the keel, or the length from the foreside of the stem to the axis of the rudder stock on that waterline, if that be greater. Where the stem contour is concave above the waterline at 85% of the least moulded depth, both the forward terminal of the total length and the foreside of the stem respectively shall be taken at the vertical projection to that waterline of the aftermost point of the stem contour (above that waterline). In ships designed with a rake of keel the waterline on which this length is measured shall be parallel to the designed waterline.* (IMO 1966). This definition is still in use today and is illustrated in Fig. 1 (Djupvik et al. [8]).

The transverse damage extent is determined by Eq. 2, where B is the beam of the ship. B is measured at the deepest subdivision draught (load line), from the ship side

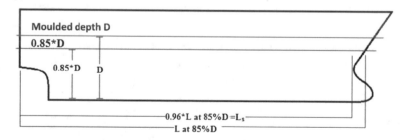

Fig. 1 Ship length as stated in ICCL-66 [8]

90° onto the centreline (Hjort and Olufsen [6]; Patterson and Ridley [7]).

$$\text{Transverse damage extent} = Min\left(\frac{B}{5}, 11.5\right) \text{[m]} \tag{2}$$

The vertical damage extent has no limitations and is taken as the entire depth of the ship. Furthermore, a worst-case loading scenario shall always be assumed. The permeability is set to, e.g., 95% for accommodation and 85% for machinery spaces. If a lesser damage causes a worse condition, then the worst case shall be used (Patterson and Ridley [7]).

As an example, the triangle illustrates a damage opening the rooms in zone 2 to the sea and the parallelogram illustrates a damage where rooms in the zones 4, 5 and 6 are flooded simultaneously.

2.2 Requirements

After the ship has reached an equilibrium position in a damage scenario where one or more compartments are exposed to flooding, and the lost buoyancy method is used to calculate the damaged trim and stability, the following requirements must be met [7], see Fig. 2:

- $GM > 0.05\,\text{m}$
- Heel angle $\leq 7°$ for one compartment flooding, or $12°$ for two or more adjacent compartments.
- Specific minimum requirements related to the area under the GZ curve (see Fig. 2).
- Range of Stability $\geq 15°$. This requirement may be reduced from $15°$ to $10°$ if the area under the GZ curve increases by a certain ratio.

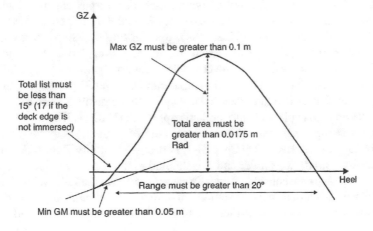

Fig. 2 Damage stability requirements pre-2009

- Peak GZ value = max ($\frac{\text{Heeling moment}}{\Delta} + 0.04, 0.10$) where the heeling moment is generated by:

  - All passengers crowding to deck areas on one side of the ship where the muster stations are located, with a passenger weight of 75 kg and a density of 4 passengers to a square metre.
  - Davit launching of all fully loaded survival crafts on one side of the ship.
  - Wind pressure = $120 \, \text{N/m}^2$ on one side of the ship.

For any type of passenger vessel, the margin line must not be submerged in the final equilibrium position.

2.3 Probabilistic Damage Stability

What is a probabilistic approach? First of all a degree of uncertainty. Thus, *random variables* are required to develop prediction models, which for example can be used to describe the behaviour of a system. There is no universal definition of *randomness*, but in the context of damage stability it is taken to mean that accidents and the damage extent of accidents are unpredictable. In order to map the unpredictable, the only available analytical tool is probability theory. Past knowledge, e.g. damage statistics, can be used to predict random factors that influence the final consequence of damage to a ships hull. Such random factors may be the mass and the velocity of the ramming ship. The influence of these random factors is different for ships with different characteristics; for instance, differences in the range of permeability and service draught (IMO [4]; Kirchsteiger [5]). The PS regulations that entered into force on the January 1 2009 as a part of SOLAS 2009 Chap. II-1, Part $B-1$ Stability, apply to dry cargo ships with a length of 80 m or more and all passenger ships with a keel laying on or after this date. A passenger ship is defined to be from the IMO definition of a ship carrying more than 12 passengers. In addition, because the *Code of Safety for Special Purpose Ships, 2008* (SPS code) was adopted in 2008 by IMO Resolution MSC.266(84), Special Purpose Ships (SPS) are also covered. Furthermore, all the ships applicable to the PS regulations are required to have a double bottom and automatic cross-flooding arrangements that stabilise the ship within 10 min. On top of this, if the ship is carrying over 36 passengers, there are additional deterministic requirements. These will not be detailed here, as the scope of this section is to explain the probabilistic approach (IMO [1,3]).

Whether a ship is *safe enough* according to the PS regulations or not, it is determined by Eq. 3. In SOLAS 2009 Chap. II-1, Part $B-1$ Stability, Reg. 7, A is defined as the *attained subdivision index* and R is defined as the *required subdivision index*. Two different ships are considered equally safe if they have the same value of A. The calculation of A is based on the probability of damage, i.e. flooding of compartments, and the survivability of the ship after flooding. This and more are explained in detail throughout this section (IMO [1]; IMO [4]; Patterson and Ridley [7]).

$$A > R \tag{3}$$

2.3.1 Limitations

The current PS regulations are based on damage statistics—more precisely, collision statistics. A collision may be ship-to-ship or contact between a ship and an obstacle, e.g. an ice berg. For assessment of groundings there is no probabilistic approach available in the regulations, most likely due to lack of reporting grounding statistics. However, a widely used technique for probabilistic approaches when relevant statistics are insufficient is the Monte Carlo simulation. In other words, one could possibly use the damage statistics from the GOALDS project as a basis in combination with Monte Carlo simulation to develop an approach. The probability distributions from GOALDS can be found in IMO-SLF 55/INF.7 (Hjort and Olufsen [6]).

Furthermore, Table 3 provides a complete overview of which ship types that follow the PS approach and which ship types that follow the DS approach.

Table 3 Overview of damage stability conventions for different ship types

Code or convention	Ship type	Method
SOLAS-2009	All passenger ships - Pure passenger ships - Ro-Ro ships - Cruise ships	Probabilistic
SPS Code/SOLAS-2009	- Special purpose ships	
SOLAS-2009	Dry cargo ships >80 m in length - RoRo cargo ships - Car carriers - General cargo ships - Bulk carriers with reduced freeboard and deck cargo (IACS unified interpretation no. 65) - Cable laying ships	Probabilistic
1966 load line convention	Dry cargo ships with reduced freeboard	Deterministic
1966 load line convention/MARPOL 73/78 Annex 1	Oil tankers	Deterministic
International bulk chemical code	Chemical tankers	Deterministic
International liquefied gas carrier code	Liquefied gas carriers	Deterministic

2.4 Probabilistic Concept

2.4.1 Background

The R-index for passenger ships was established based on sample ship calculations from the HARDER project (Hjort and Olufsen [6]). More information on the development of the R-index is difficult to obtain.

An important objective of the development of PS regulations was to ensure that new and existing ships should have approximately the same level of safety with the PS regulations. Bearing this in mind, an initial formula for the R-index for passenger ships was developed in a scientific manner, by carrying out multiple test runs using existing ships. The problem with this approach was that the value of the R-index declined with increasing values of ship length and number of passengers. This was totally unacceptable for the membership countries of IMO; it was argued that it would give an unbalanced picture of the safety level of large existing passenger ships. In consequence, a *political correct* compromise was agreed upon and a new, but not necessarily scientific *correct* formula for the R-index was developed. The explanation of the declining value of R with the initial formula for passenger ships is not clear, but some thoughts are summarised here:

With the old, deterministic rules the safety standard of the ship depended on the degree of watertight subdivision of the ship, i.e. the number of watertight bulkheads fitted. The damage extent represented the distance between watertight bulkheads, which was maximum 11 m. Thus, for the larger ships that were defined as two compartment ships by means of damage stability, the damage extent was 22 m. In comparison, the maximum damage extent in the PS regulations is 60 m. In other words, the relative damage extent was very low for the large ships and, thus, one may expect a declining R-index.

The deterministic *B/5 rule*, of which the watertight arrangements for existing ships at the time usually were optimised against, might also have had an impact on the unexpected declining R-index value; in the development of the probabilistic rules, the statistics from the HARDER project showed that the maximum transverse damage extent should be $B/2$, not $B/5$.

The new SOLAS 2009 regulations applies for dry cargo ships of 80 m in length (L) and upwards and to all passenger vessels with a keel laying date on or after 1 January 2009, respectively, for vessels which undergo a major conversion on or after that date. The harmonised regulations on subdivision and damage stability are contained in SOLAS Chap. II-1 in parts $B - 1$ through to $B - 4$. These regulations use the probability of survival after damage as a measure of ships' safety in a damaged condition. They are intended to provide ships with a minimum standard of subdivision determined by the required subdivision index R which depends on the ship length and number of passengers. Any assumed damage of arbitrary extent can make a contribution towards establishing of the required subdivision index R. The probability to survive in each case of damage is assessed, and the summation of all positive probabilities of survival provides an attained subdivision index A. The attained subdivision index A is to be not less than the required subdivision index R.

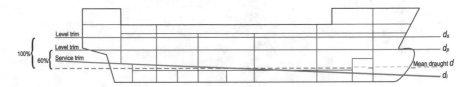

Fig. 3 Waterlines used in probabilistic damage calculations [4]

For cargo ships, the R-index formula was based on a probabilistic approach from the beginning, so there has been no major changes affecting the results. This in turn gives quite equal results for new and old cargo ships. However, it was commonly accepted that some designs would deviate due to the new foundation of statistical data, for instance from the HARDER project. It should also be noted that the old rules concerning dry cargo ships in reality were a compromise built on the *any rules are better than no rules* mentality. In the development of the R-index, or the PS regulations in general, they had to deal with these rules. As mentioned before, politics has always played an important role in the development of IMO regulations, and formulas may be developed based on political compromises. As a result, it can be difficult to understand the formulae completely.

Calculations are to be carried out for three initial draughts, see Fig. 3:

- a deepest subdivision draught without trim, d_s,
- a partial subdivision draught without trim, d_p,
- a light service draught with a trim level corresponding to that condition, d_l.

For each of the three considered draughts the calculated partial index A_s, A_p and A_l shall meet a percentage of the total attained index A. For dry cargo vessels that percentage shall be not less than $0.5R$, for passenger vessels $0.9R$.

$$A = 0.4A_s + 0.4A_p + 0.2A_l \qquad (4)$$

2.4.2 Zone Damage

In order to prepare the calculation of A, the ship under consideration must be divided into a fixed discrete number of zones, in longitudinal, transverse and vertical direction. A longitudinal zone, or just *zone*, is defined as *a longitudinal interval of the ship within the subdivision length* (IMO [4]). It is up to the designer how the zone division is done; the only rule for subdivision is that the subdivision length l_s defines the extremes for the hull in longitudinal direction, as shown in Figs. 5, 6 and 7. In order to maximise safety, the goal should be to obtain as large A-index as possible. Thus, it is important to be strategic when doing the subdivision, since each zone and all combinations of adjacent zones contribute to the A-index. More zones do in general give a larger A-index. The number of zones should however be limited to some extent, in order to keep the computation time at an acceptable level. One strategy may be to divide the zones according to the watertight subdivision of the ship,

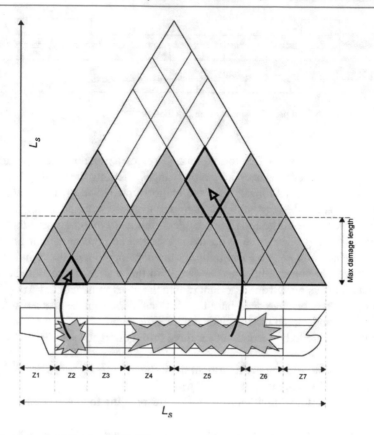

Fig. 4 Example of subdivision (IMO [4])

which is said to give profitable results. Furthermore, Fig. 4 illustrates a seven-zone division of a ship with the corresponding possible single- and multi-zone damages. The bottom line triangles indicate single-zone damages, whilst the parallelograms indicate multi-zone damages IMO [1] (Djupvik et al. [8]; IMO [4]; Lützen [9]).

2.5 Excerpt from ANNEX 22 of SOLAS

The harmonised SOLAS regulations on subdivision and damage stability, as contained in SOLAS Chap. II-1, are based on a probabilistic concept which uses the probability of survival after collision as a measure of ships' safety in a damaged condition. This probability is referred to as the *attained subdivision index A* in the regulations. The probability that a ship will remain afloat without sinking or capsizing as a result of an arbitrary collision in a given longitudinal position can be broken down into:

- The probability that the longitudinal centre of damage occurs in just the region of the ship under consideration:
- The probability that this damage has a longitudinal extent that only includes spaces between the transverse watertight bulkheads found in this region;
- The probability that the damage has a vertical extent that will flood only the spaces below a given horizontal boundary, such as a watertight deck;
- The probability that the damage has a transverse penetration not greater than the distance to a given longitudinal boundary; and
- The probability that the watertight integrity and the stability throughout the flooding sequence is sufficient to avoid capsizing or sinking.

The first three of these factors are solely dependent on the watertight arrangement of the ship, whilst the last two depend on the ship's shape. The last factor also depends on the actual loading condition. By grouping these probabilities, calculations of the probability of survival, or attained index A, have been formulated to include the following probabilities:

- The probability of flooding each single compartment and each possible group of two or more adjacent compartments; and
- The probability that the stability after flooding a compartment or a group of two or more adjacent compartments will be sufficient to prevent capsizing or dangerous heeling due to loss of stability or to heeling moments in intermediate or final stages of flooding.

This concept allows a rule requirement to be applied by requiring a minimum value of A for a particular ship. This minimum value is referred to as the required subdivision index R in the present regulations and can be made dependent on ship size, number of passengers or other factors legislators might consider important.

Evidence of compliance with the rules then simply becomes:

$$A \geq R \qquad (5)$$

As explained above, the attained subdivision index A is determined by a formula for the entire probability as the sum of the products for each compartment or group of compartments of the probability that a space is flooded, multiplied by the probability that the ship will not capsize or sink due to flooding of the considered space. In other words, the general formula for the attained index can be given in the form:

$$A = \sum p_i s_i \qquad (6)$$

Subscript i represents the damage zone (group of compartments) under consideration within the watertight subdivision of the ship. The subdivision is viewed in the longitudinal direction, starting with the aftmost zone/compartment. The value of p_i represents the probability that only the zone i under consideration will be flooded, disregarding any horizontal subdivision, but taking transverse subdivision into account. Longitudinal subdivision within the zone will result in additional flooding scenarios, each with its own probability of occurrence. The value of i represents the probability of survival after flooding the zone i under consideration.

In regulation $7-1$, the words compartment and group of compartments should be understood to mean zone and adjacent zones. Zone is defined as a longitudinal interval of the ship within the subdivision length. Room is defined as a part of the ship, limited by bulkheads and decks, having a specific permeability. Space is a combination of rooms. Compartment is an onboard space within watertight boundaries. Damage is the three-dimensional extent of the breach in the ship. For the calculation of p, v, r and b only the damage should be considered, and for the calculation of the s-value the flooded space should be considered.

Although the ideas outlined above are very simple, their practical application in an exact manner would give rise to several difficulties if a mathematically perfect method was to be developed. As pointed out above, an extensive but still incomplete description of the damage will include its longitudinal and vertical location as well as its longitudinal, vertical and transverse extent. Apart from the difficulties in handling such a five-dimensional random variable, it is impossible to determine its probability distribution very accurately with the presently available damage statistics. Similar limitations are true for the variables and physical relationships involved in the calculation of the probability that a ship will not capsize or sink during intermediate stages or in the final stage of flooding.

A close approximation of the available statistics would result in extremely numerous and complicated computations. In order to make the concept practicable, extensive simplifications are necessary. Although it is not possible to calculate the exact probability of survival on such a simplified basis, it has still been possible to develop a useful comparative measure of the merits of the longitudinal, transverse and horizontal subdivision of a ship.

3   Detailed Regulations According to SOLAS 2009

3.1   Subdivision Length

Before explaining any further how the R and A indexes are calculated, it is useful to introduce a frequently used factor named *subdivision length*, which is denoted l_s in the PS regulations. It is important to distinguish between this length factor and the one used in the deterministic damage stability (DS) regulations. Figures 5, 6 and 7 illustrate how the subdivision length is determined for three scenarios. As these figures show, the subdivision length depends on the buoyant hull and the reserve buoyancy of the ship, and whether these *areas* are damaged or not. The buoyant hull comprises the enclosed volume of the ship below the waterline, which is denoted d_s in the figures, whilst the reserve buoyancy is comprised by the enclosed volume of the ship above the waterline. The black line is defined as the maximum vertical damage extent and is always equal to $d_s + 12.5$ m measured from the baseline. The

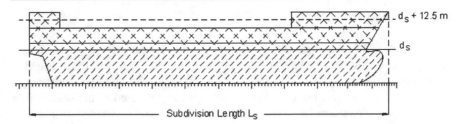

Fig. 5 Example 1 of how the subdivision length is determined

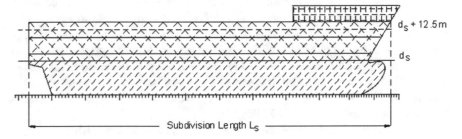

Fig. 6 Example 2 of how the subdivision length is determined

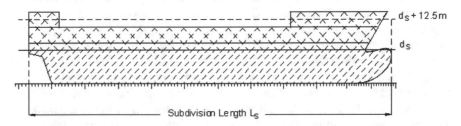

Fig. 7 Example 3 of how the subdivision length is determined

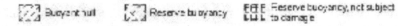

Fig. 8 Legend for Figs. 5, 6 and 7

ship illustrated in Fig. 7 distinguishes between reserve buoyancy that is harmed and unharmed. Fig. 8 shows the legend used in Figs 5, 6 and 7. The subdivision length is then measured from the stern to the foremost point of the harmed area at the stem.

The limiting deck for the reserve buoyancy may be partially watertight. The maximum possible vertical extent of damage above the baseline is $d_s + 12.5\,\text{m}$ (Fig. 8).

3.2 Calculation Method

The outcome of DS calculations, i.e. whether the ship is *safe enough* or not, mainly depends on the ship size; the length, beam and depth of the ship in question. Together these three parameters determine the damage extent of the ship, as explained in previous chapters of this book. In general, the parameters used in the calculations are the same for different ship types, but the impact of the parameters differ. This impact depends on factors such as ship type, ship size, cargo type and number of passengers.

Furthermore, the ship in question is required to survive certain damage scenarios, given by the predetermined damage extent. The goal is of course to identify the most critical damage scenarios. This is done by investigating all possible damage conditions within the boundaries of the damage extent. All damage scenarios must comply with the requirements as defined in SOLAS. Some of these requirements are presented in the previous section. If the results from the calculations are not up to the standard, the ship will not be approved by the flag state or the classification societies (Djupvik et al. [8]; Patterson and Ridley [7]).

The harmonised SOLAS regulations on subdivision and damage stability, as contained in SOLAS Chap. II-1, are based on a probabilistic concept which uses the probability of survival after collision as a measure of ships' safety in a damaged condition. This probability is referred to as the attained subdivision index *A* in the regulations. It can be considered an objective measure of ships' safety and, ideally, there would be no need to supplement this index by any deterministic requirements. The philosophy behind the probabilistic concept is that two different ships with the same attained index are of equal safety and, therefore, there is no need for special treatment of specific parts of the ship, even if they are able to survive different damages. The only areas which are given special attention in the regulations are the forward and bottom regions, which are dealt with by special subdivision rules provided for cases of ramming and grounding.

Only a few deterministic elements, which were necessary to make the concept practicable, have been included. It was also necessary to include a deterministic minor damage on top of the probabilistic regulations for passenger ships to avoid ships being designed with what might be perceived as unacceptably vulnerable spots in some part of their length.

It is easily recognised that there are many factors that will affect the final consequences of hull damage to a ship. These factors are random, and their influence is different for ships with different characteristics. For example, it would seem obvious that in ships of similar size carrying different amounts of cargo, damage of similar extent may lead to different results because of differences in the range of permeability and draught during service. The mass and velocity of the ramming ship is obviously another random variable.

Due to this, the effect of a three-dimensional damage to a ship with given watertight subdivision depends on the following circumstances:

- which particular space or group of adjacent spaces is flooded;
- the draught, trim and intact metacentric height at the time of damage;
- the permeability of affected spaces at the time of damage;
- the sea state at the time of damage; and
- other factors such as possible heeling moments due to unsymmetrical weights.

Some of these circumstances are interdependent, and the relationship between them and their effects may vary in different cases. Additionally, the effect of hull strength on penetration will obviously have some effect on the results for a given ship. Since the location and size of the damage is random, it is not possible to state which part of the ship becomes flooded. However, the probability of flooding a given space can be determined if the probability of occurrence of certain damages is known from experience, that is damage statistics. The probability of flooding a space is then equal to the probability of occurrence of all such damages which just open the considered space to the sea.

For these reasons and because of mathematical complexity as well as insufficient data, it would not be practicable to make an exact or direct assessment of their effect on the probability that a particular ship will survive a random damage if it occurs. However, accepting some approximations or qualitative judgments, a logical treatment may be achieved by using the probability approach as the basis for a comparative method for the assessment and regulation of ship safety. It may be demonstrated by means of probability theory that the probability of ship survival should be calculated as the sum of probabilities of its survival after flooding each single compartment, each group of two, three, etc., adjacent compartments multiplied, respectively, by the probabilities of occurrence of such damages leading to the flooding of the corresponding compartment or group of compartments.

3.3   Longitudinal Subdivision

In order to prepare for the calculation of index A, the ship's subdivision length l_s is divided into a fixed discrete number of damage zones. These damage zones will determine the damage stability investigation in the way of specific damages to be calculated. There are no rules for the subdividing, except that the length l_s defines the extremes for the actual hull. Zone boundaries need not coincide with physical watertight boundaries. However, it is important to consider a strategy carefully to obtain a good result (i.e. a large attained index A). All zones and combination of adjacent zones may contribute to the index A. In general it is expected that the more zone boundaries the ship is divided into the higher will be the attained index, but this benefit should be balanced against extra computing time. Figure 9 shows different longitudinal zone divisions of the length l_s.

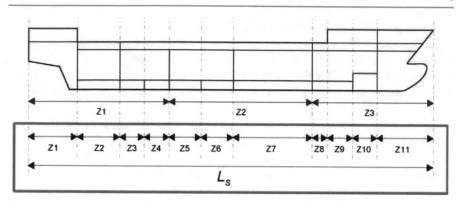

Fig. 9 Examples of how the subdivision length is determined

3.4 Regulation Definitions

If a passenger ship built before 1 January 2009 undergoes alterations or modifications of major character, it may still remain under the damage stability regulations applicable to ships built before 1 January 2009, except in the case of a cargo ship being converted to a passenger ship.

3.5 Light Service Draught

The light service draught (d_l) represents the lower draught limit of the minimum required GM (or maximum allowable KG) curve. It corresponds, in general, to the ballast arrival condition with 10% consumables for cargo ships. For passenger ships, it corresponds, in general, to the arrival condition with 10% consumables, a full complement of passengers and crew and their effects, and ballast as necessary for stability and trim. The 10% arrival condition is not necessarily the specific condition that should be used for all ships, but represents, in general, a suitable lower limit for all loading conditions. This is understood to not include docking conditions or other non-voyage conditions.

3.6 Draught and Trim

Linear interpolation of the limiting values between the draughts d_s, d_p and d_l is only applicable with due respect to minima GM values. If it is intended to develop curves of maximum permissible KG, a sufficient number of KM_T values for intermediate draughts should be calculated to ensure that the resulting maximum KG curves correspond with a linear variation of GM. When light service draught is not with the same trim as other draughts, KM_T for draughts between partial and light service draught should be calculated for trims interpolated between trim at partial draught and

Fig. 10 Effects of draught on *GM* (IMO [4])

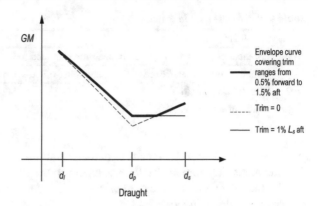

trim at light service draught. In cases where the operational trim range is intended to exceed $\pm 0.5\%$ of l_s, the original *GM* limit line should be designed in the usual manner with the deepest subdivision draught and partial subdivision draught calculated at level trim and actual service trim used for the light service draught. Then additional sets of *GM* limit lines should be constructed on the basis of the operational range of trims which is covered by loading conditions of partial subdivision draught and deepest subdivision draught ensuring that intervals of 1% l_s are not exceeded. For the light service draught d_l only one trim should be considered. The sets of *GM* limit lines are combined to give one envelope limiting *GM* curve. The effective trim range of the curve should be clearly stated see Fig. 10.

3.7   Required Subdivision Index R

The *R*-index for passenger ships was established based on sample ship calculations from the HARDER project (Hjort and Olufsen [6]). An important objective of the development of PS regulations was to ensure that new and existing ships should have approximately the same level of safety with the PS regulations. Bearing this in mind, an initial formula for the *R*-index for passenger ships was developed in a scientific manner, by carrying out multiple test runs using existing ships. The problem with this approach was that the value of the *R*-index declined with increasing values of ship length and number of passengers. This was totally unacceptable for the membership countries of IMO; it was argued that it would give an unbalanced picture of the safety level of large existing passenger ships. In consequence, a *political correct* compromise was agreed upon and a new, but not necessarily scientific *correct* formula for the *R*-index was developed. The explanation of the declining value of *R* with the initial formula for passenger ships is not clear, but some thoughts are summarised here:

   With the old, deterministic rules the safety standard of the ship depended on the degree of watertight subdivision of the ship, i.e. the number of watertight bulkheads fitted. The damage extent represented the distance between watertight bulkheads, which was maximum 11 m. Thus, for the larger ships that were defined as

Table 4 Parameters used in R index

N	$N_1 + 2N_2$
N_1	Number of persons for whom lifeboats are provided
N_2	Number of persons in excess of N_1, including officers and crew
l_s	Subdivision length
R_0	Value of R from Eq. 9

two compartment ships by means of damage stability, the damage extent was 22 m. In comparison, the maximum damage extent in the PS regulations is 60 m. In other words, the relative damage extent was very low for the large ships and, thus, one may expect a declining R-index.

The deterministic *B/5 rule*, of which the watertight arrangements for existing ships at the time usually were optimised against, might also have had an impact on the unexpected declining R-index

The calculation procedure for the required subdivision index R (R-index) is depen-dent on the ship type. For passenger ships, the R-index is a function of ship length, number of persons on board and the lifeboat capacity, as shown by Eq. 7. For cargo ships, the R-index is solely a function of the ship length. For cargo ships larger than 100 m in length and cargo ships between 80 and 100 m in length, R-index is calculated by Eqs. 8 and 9, respectively.

Explanations to the parameters used in the below equations are gathered in Table 4.

$$R = \frac{5000}{l_s + 2.5N + 15225} \tag{7}$$

$$R = 1 - \frac{128}{l_s + 152} \tag{8}$$

$$R = 1 - \left[\frac{1}{1 + \frac{l_s}{100} \frac{R_0}{1 - R_0}} \right] \tag{9}$$

Regarding the term's reduced degree of hazard, the following interpretation should be applied: A lesser value of N, but in no case less than $N = N_1 + N_2$, may be allowed at the discretion of the administration for passenger ships, which, in the course of their voyages, do not proceed more than 20 miles from the nearest land.

For SPSs, there are some other specific requirements. In general, the SPS is con-sidered a passenger ship in accordance with SOLAS Chap. II-1, and special personnel are considered passengers. However, the requirements related to the R-index varies with the number of persons that the SPS is allowed to carry (IMO [3]):

- the R value is taken as R if the SPS is certified to carry 240 persons or more;
- the R value is taken as $0.8R$ if the SPS is certified to carry not more than 60 persons;

- the R value is determined by linear interpolation between the R values given in 1 and 2 above, if the SPS is carrying more than 60 but less than 240 persons.

3.8 Attained Subdivision Index A

The probability of surviving after collision damage to the ship's hull is expressed by the index A. Producing an index A requires calculation of various damage scenarios defined by the extent of damage and the initial loading conditions of the ship before damage. Three loading conditions should be considered, and the result weighted as follows:

$$A = 0.4A_s + 0.4A_p + 0.2A_l$$

The method for calculating A for a loading condition is expressed by the formula:

$$A_c = \sum_{i=1}^{i=t} p_i s_i v_i \tag{10}$$

The index c represents one of the three loading conditions, the index i represents each investigated damage or group of damages, and t is the number of damages to be investigated to calculate A_c for the particular loading condition.

To obtain a maximum index A for a given subdivision, t has to be equal to T, the total number of damages. In practice, the damage combinations to be considered are limited either by significantly reduced contributions to A (i.e. flooding of substantially larger volumes) or by exceeding the maximum possible damage length.

The index A is divided into partial factors as follows:

p_i The p factor is solely dependent on the geometry of the watertight arrangement of the ship.

s_i The s factor is dependent on the calculated survivability of the ship after the considered damage for a specific initial condition.

v_i The v factor is dependent on the geometry of the watertight arrangement (decks) of the ship and the draught of the initial loading condition. It represents the probability that the spaces above the horizontal subdivision will not be flooded.

Three initial loading conditions should be used for calculating the index A. The loading conditions are defined by their mean draught d, trim and GM (or KG). The mean draught and trim are illustrated in Fig. 3.

The GM (or KG) values for the three loading conditions could, as a first attempt, be taken from the intact stability GM (or KG) limit curve. If the required index R is not obtained, the GM (or KG) values may be increased (or reduced), implying that the intact loading conditions from the intact stability book must now meet the GM (or KG) limit curve from the damage stability calculations derived by linear interpolation between the three GMs.

Table 5 Parameters used in p_i index

j	The aftmost zone number involved in the damage starting with No.1 at the stern
n	The number of adjacent damage zones involved in the damage
k	The number of a particular longitudinal bulkhead as barrier for the transverse penetration in the damage zone counted from shell towards the centreline. The shell has $k = 0$
$x1$	The distance from the aft terminal of l_s to the after end of the zone in question
$x2$	The distance from the aft terminal of l_s to the forward end of the zone in question
b	The mean transverse distance in metres measured at right angles to the centreline at the deepest subdivision load line between the shell and an assumed vertical plane extended between the longitudinal limits with, all or part of the outermost portion of the longitudinal bulkhead under consideration

3.8.1 Calculation of the p_i Factor

The p_i factor depends upon the watertight arrangements and the zone division of the ship. This factor is for the probability of specific damage to the ship, without any horizontal subdivision. The vertical damage extent is accounted for in the v_i factor. For single-zone damage then use Eq. 12. It must be recalled that

$$\sum p_i = 1 \tag{11}$$

If the damage involves just a single zone then

$$p_i = p(x1_j, x2_j) \times [r(x1_j, x2_j, b_k) - r(x1_j, x2_j, b_{k-1})] \tag{12}$$

If the damage involves two adjacent zones then

$$\begin{aligned} p_i = \quad & p(x1_j, x2_{j+1}) \times [r(x1_j, x2_{j+1}, b_k) - r(x1_j, x2_{j+1}j, b_{k-1})] \\ & - p(x1_j, x2_j) \times [r(x1_j, x2_j, b_k) - r(x1_j, x2_j, b_{k-1})] \\ - & p(x1_{j+1}, x2_{j+1}) \times [r(x1_{j+1}, x2_{j+1}, b_k) - r(x1_{j+1}, x2_{j+1}, b_{k-1})] \end{aligned} \tag{13}$$

There are equivalent equations when the damage is for three adjacent compartments. According to the Appendix A in [1] where all the parameters are given in Table 5, and in particular

$$r(x1, x2, b_0) = 0 \tag{14}$$

The factor $p(x1, x2)$ is to be calculated according to the following formulae:

Overall normalised maximum damage length $J_{max} = \frac{10}{33}$

Knuckle point in the distribution $\qquad J_{kn} = \frac{5}{33}$

Cumulative probability at J_{kn} $\qquad p_k = \frac{11}{12}$

Maximum absolute damage length $\qquad l_{max} = 60\,\text{m}$

Length where normalised distribution ends $\quad L^* = 260\,\text{m}$

The probability density at $J = 0$;

$$b_o = 2 \left[\frac{p_k}{J_{kn}} - \frac{1 - p_k}{J_{max} - J_{kn}} \right] \tag{15}$$

Now when $l_s \leq L^*$:

$$J_m = \min \left(J_{max}, \frac{l_{max}}{l_s} \right) \tag{16}$$

$$J_k = \frac{J_m}{2} + \frac{1 - \sqrt{1 + (1 - 2p_k)b_o J_m + \frac{1}{4}b_0^2 J_m^2}}{b_0} \tag{17}$$

$$b_{12} = b_0 \tag{18}$$

When $l_s \geq L^*$,

$$J_m^* = \min \left(J_{max}, \frac{l_{max}}{L^*} \right) \tag{19}$$

$$J_k^* = \frac{J_m^*}{2} + \frac{1 - \sqrt{1 + (1 - 2p_k)b_o J_m^* + \frac{1}{4}b_0^2 J_m^{*2}}}{b_0} \tag{20}$$

$$J_m = \frac{J_m^* L^*}{l_s} \tag{21}$$

$$J_k = \frac{J_k^* L^*}{l_s} \tag{22}$$

$$b_{12} = 2 \left(\frac{p_k}{J_k} - \frac{1 - p_k}{J_m - J_k} \right) \tag{23}$$

$$b_{11} = 4 \frac{1 - p_k}{(J_m - J_k)J_k} - 2\frac{p_k}{J_k^2} \tag{24}$$

$$b_{21} = -2\frac{1 - p_k}{(J_m - J_k)^2} \tag{25}$$

$$b_{22} = -b_{21}J_m \tag{26}$$

The non-dimensional damage length is:

$$J = \frac{x2 - x1}{l_s} \tag{27}$$

The normalised length of a compartment or group of compartments is:

$$J_n = \min (J, J_m) \tag{28}$$

where neither limits of the compartment or group of compartments under question coincides with the aft or forward terminals, then for $J = J_k$

$$p(x1, x2) = p_1 = \frac{1}{6}J^2(b + 11J + 3b_{12}) \tag{29}$$

and if $J > J_k$ then

$$p(x1, x2) = p_2 = \frac{1}{3}J_k^2 + \frac{1}{2}(b_{11}J - b_{12})J_k^2 + b_{12}J J_k - \frac{1}{3}(J_n^3 - J_k^3)$$
$$+ \frac{1}{2}(b_{21}J - b_{22})(J_n^2 - J_k^2) + b_{22}J(J_n - J_k) \tag{30}$$

where the aft limit of the compartment or group of compartments under question coincides with the aft terminal or the forward limit of the compartment or group of compartments coincides with the forward terminal then for $J = J_k$

$$p(x1, x2) = \frac{1}{2}(p_1 + J) \tag{31}$$

If $J > J_k$ then

$$p(x1, x2) = \frac{1}{2}(p_2 + J) \tag{32}$$

where the compartment or group of compartments under question extends over the entire subdivision length l_s, then:

$$p(x_1, x2) = 1 \tag{33}$$

The definition is

$$r(x1, x2, b) = 1 - (1 - C)\left[1 - \frac{G}{p(x1, x2)}\right] \tag{34}$$

with

$$C = 12J_b(-45J_b + 4) \tag{35}$$

and,

$$J_b = \frac{b}{15B} \tag{36}$$

where the definition of G is evaluated from,

$$G = G_1 = \frac{1}{2}b_{11}J_b^2 + b_{12}J_b \tag{37}$$

where compartment or groups extend over the entire L_s, or

$$G = G_2 = -\frac{1}{3}b_{11}J_0^3 + \frac{1}{2}(b_{11}J - b_{12})J_0^2 + b_{12}J J_0 \tag{38}$$

where neither end of compartment is aft or forward terminals and

$$J_0 = \min (J, J_b) \tag{39}$$

If the compartment or group coincides with either terminals then

$$G = \frac{1}{2}(G_2 + G_1 J) \tag{40}$$

3.8.2 Calculation of the s_i Factor

A number of criteria in the PS regulations concerning the s_i factor appear quite deterministic. The waterline concept of the DS regulations covers the issues related to watertight integrity, ship safety and the evacuation of passengers and crew. There is a difference between PS and DS methodology regarding passenger ships in that the margin line requirement is removed. The concept of the margin line is to prohibit water immersion on the bulkhead deck, and this is considered concept. The effect had to compensated for by some explicit requirements considering watertight integrity and evacuation of passengers. Examples include the life-saving equipment. Another important compromise was the introduction of penalties on the A index. These penalties were given in the event water immersion of certain features such as water on the bulkhead deck during evacuation, immersion of vertical emergency escape hatches and progressive flooding through unprotected openings or damaged piping and ducts in the damaged zone. It must be recalled that maximum vertical damage is defined as $d + 12.5\,\mathrm{m}$.

The factor s_i can be determined for each case of flooding involving a single or group of compartments. This requires the following defined variables:

θ_e is the equilibrium heel angle in any stage of flooding in degrees,
θ_v is the angle in any stage of flooding where the righting lever becomes positive or negative, or the angle at which an opening incapable of being closed watertight becomes submerged,
GZ_{max} is the maximum positive righting lever, in metres, up to the angle θ_v,
Range is the range of positive righting levers in degrees, measured from angle θ_e, and the positive range is to be taken up to the angle θ_v,
Flooding stage is the discrete step during the flooding process, including the stage before equalisation until final equilibrium.

The factor s_i for any damage case at any flooding condition, d_s, shall be obtained from:

$$s_i = \min (s_{intermediate,i}, s_{final,i} s_{mom,i}) \tag{41}$$

where the variables are defined as:

$s_{intermediate,i}$ is the probability to survive all intermediate flooding stages until the
final equilibrium stage and is calculated as in Eq. 42
$s_{final,i}$ is the probability to survive in the final equilibrium stage of flooding using
Eq. 43
$s_{mom,i}$ is the probability to survive heeling moments using Eq. 44

The factor $s_{intermediate,i}$ is applicable only for passenger ships (for cargo ships
$s_{intermediate,i}$ should be taken as unity) and shall be the least of the s factors obtained
from all flooding stages including the stage before equalisation, if any, and is calculated as:

$$s_{intermediate,i} = \left[\frac{GZ_{max}}{0.05} \frac{Range}{7} \right]^{\frac{1}{4}} \tag{42}$$

where GZ_{max} is not to be taken as more than 0.05 m and $Range$ is not more than 7°.
$s_{intermediate,i} = 0$, if the intermediate heel angle exceeds 15°. Where cross-flooding
fittings are required, the time for equalisation shall not exceed 10 min.
The factor, $s_{final,i}$ shall be obtained from:

$$s_{final,i} = K \left[\frac{GZ_{max}}{0.12} \frac{Range}{16} \right]^{\frac{1}{4}} \tag{43}$$

where the variables used in Eq. 43 are:

GZ_{max} is not to be taken as more than 0.12 m
$Range$ is not to be taken as more than 16°
$K = 1$ if $\theta_e \leq \theta_{min}$
$K = 0$ if $\theta_e \geq \theta_{min}$
$K = \sqrt{\frac{\theta_{max} - \theta_e}{\theta_{max} - \theta_{min}}}$ otherwise

and, θ_{min} is 7° for passenger ships and 25° for cargo ships. Factor θmax is 15° for
passenger ships and 30° for cargo ships.
The factor $s_{mom,i}$ is applicable only to passenger ships (for cargos ships $s_{mom,i}$
shall be taken as unity) and shall be calculated at the final equilibrium using:

$$s_{mom,i} = \frac{(GZ_{max} - 0.04)Displacement}{M_{heel}} \tag{44}$$

where the following definitions apply:

$Displacement$ is the intact displacement at the subdivision draught
M_{heel} is the maximum assumed heeling moment as in Eq. 45
$s_{mom,i}$ ≤ 1

The wind heeling moment M_{heel} is calculated from:

$$M_{heel} = \max (M_{passenger}, M_{wind}, M_{survivalcraft}) \tag{45}$$

The moment definitions are:

$$M_{passenger} = (0.075 N_p)(0.45 B) \tag{46}$$

where N_p is the maximum number of passengers permitted to be on board in the service condition corresponding to the deepest subdivision draught under consideration and B is the ship beam.

M_{wind} is the maximum assumed wind force acting in a damage situation:

$$M_{wind} = \frac{(PAZ)}{9.806} \tag{47}$$

where the terms in Eq. 47 are:

P 120 N/m^2
A projected lateral area above the waterline (m^2)
Z distance from centre of lateral projected area above waterline $t/2$, (m)
T ship's draught, d_i, (m)

$M_{survivalcraft}$ is the maximum assumed heeling moment due to launching of all fully laden davit-launched survival craft on one side of the ship. More details on the actual assumptions used here can be found in Annex A p. 29 of MSC 82/24/Add.1.

3.8.3 Calculation of the v_i Factor

In the case of any watertight subdivisions above the waterline, the s_i value for lower compartments or group of compartments must be multiplied by a reduction factor v_i. This factor accounts for the probability that the spaces above the horizontal boundary under consideration remain intact after a collision. If these spaces are flooded due to the ship collision, the residual stability of the whip will be reduced. In consequence, the ship's buoyancy changes and the GZ curve is affected adversely. The v_i factor is determined from Eq. 48.

$$v_i = v(H_{j,n,m}, d) - v(H_{j,n,m-1}, d) \tag{48}$$

where

$$v_i \varepsilon [0, 1] \tag{49}$$

$H_{j,n,m}$ is the least height above the baseline, in metres within longitudinal range of
   $x1(j) \dots x2(j+1)$ of the mth horizontal boundary which is assumed to limit the
   vertical extent of flooding.

$H_{j,n,(m-1)}$ is the least height above the baseline, in metres within longitudinal range of $x1(j)\ldots x2(j+1)$ of the $(m-1)$th horizontal boundary which is assumed to limit the vertical extent of flooding.

j the aft terminal of the damaged zone.

n the number of adjacent damage zones.

m the horizontal boundary counted upwards from the waterline

d draught under consideration.

The calculation of the terms $v(H_{j,n,m}, d)$ and $v(H_{j,n,m-1}, d)$ is defined in Eq. 50,

$$
v(H, d) = \begin{cases}
0.8\frac{(H-d)}{7.8} & : \quad \text{if } (H-d) \leq 7.8 \\
& : \\
0.8 + 0.2\frac{(H-d)-7.8}{4.7} & : \quad \text{if } (H-d) \geq 7.8 \\
1 & : \quad \text{if } H_m \text{ coincides with the uppermost boundary} \\
0 & : \quad \text{if } m = 0
\end{cases}
$$

(50)

where $v(H_{j,n,m})$, d is to be taken as 1, if H_m coincides with the aftermost watertight boundary of the ship within the range, $x1j\ldots x2_{j+n-1}$, and $v(H_{j,n,o})$, d is taken to be 0.

In no case is v_m to be taken as less than zero or more than 1.

In general each contribution to the A index in the event of horizontal subdivision is denoted by d_A, and then each is defined in Eq. 51.

$$
\delta_A = p_i[v_1 s_{min1} + (v_2 - v_1)s_{min2} + \cdots + (1 - v_{m-1})s_{minm}]
$$

(51)

where

v_m the v value calculated in accordance with the previous values.

s_{min} is the least s factor for all combinations of damage obtained when when the assumed damage extends from the assumed damage height H_m downwards.

4 Permeability

For the purposes of subdivision and damage stability calculations of the regulations, the permeability of each general cargo compartment or part of a compartment shall be as follows (Table 6):

The last permeability is chosen to be the one with the most severe requirement.

For the purposes of the subdivision and damage stability calculations of the regulations, the permeability of each cargo compartment of part of a compartment will be defined as in Table 7.

Table 6 Permeability regulations

Spaces	Permeability
Appropriate to stores	0.60
Occupied by accommodation	0.95
Occupied by machinery	0.85
Void spaces	0.95
Intended for liquids	0 or 0.95

Table 7 Permeability regulations for cargo ships

Spaces	Permeability at draught d_s	Permeability at draught d_p	Permeability at draught d_1
Dry cargo spaces	0.70	0.80	0.95
Container spaces	0.70	0.80	0.95
Ro-Ro spaces	0.90	0.90	0.95
Cargo liquids	0.70	0.80	0.95

References

1. IMO 2006 Solas Chapter II-1 construction - structure, subdivision and stability. IMO 2006
2. IMO 2008a International code on intact stability (2008 IS Code). International Maritime Organization. IMO 2008
3. IMO 2008b resolution MSC. 226(84) - Code of safety for special purpose ships. International Maritime Organization. IMO 2008
4. IMO 2008c resolution MSC. 281(85) Expanatory notes to the SOLAS Chapter II-1: subdivision and damage stability regulations. International Maritime Organization. IMO 2008
5. C. Kirchsteiger, On the use of probabilisitc and deterministic methods in risk analysis. J. Loss Prev. Process Ind. **12**(5), 399–419 (1999)
6. G. Hjort, O. Olufsen, Probabilistic damage stability: DNV-GL (2014)
7. C.J. Patterson, J.D. Ridley, *Ship Stability, Powering and Resistance*, vol. 13 (Adlard Coles Nautical, London, 2014)
8. O.M. Djupvik, S.A. Aanondsen, B.E. Asbjørnslett, Probabilisitic damage stability: maximising the attained index by analyzing the effects of changes in the arrangements for offshore vessels NTNU (2015)
9. M. Lützen, Ph.D. thesis: ship collision damage. Lyngby, Denmark (2001)

Second Generation Stability Methodology

14

1 Introduction

New intact stability criteria are being developed since the re-establishment in 2002 of the Sub-Committee on stability and Load Lines and on Fishing Vessels Safety (SLF) at the International Maritime Organization (IMO) in order to take into account five particular stability failure modes (see [1]). These are parametric rolling, pure loss of stability in astern waves, broaching-to, dead ship condition and excessive acceleration. For each failure mode, three levels are defined to assess the ship vulnerability with a gradually increasing level of accuracy in the prediction of ship response. The first level aims to be the more conservative and applicable with simple means such as pocket calculator. The second level could require the use of Excel spreadsheet or coding software while the third one consists of Direct Stability Assessment (DSA) carried out through the use of software implementing state-of-the-art ship dynamics models. We might also observe a rather static analysis at the first level evolving into a totally dynamic analysis at the third level. A ship is required to comply with one of the three criteria for each failure mode. If a vessel is considered as vulnerable according to the first level, then the second level check is carried out. A vessel failing the second level check must be subject to a DSA. In the case of a vessel being still considered vulnerable, operational guidance and/or limitations should be applied. Figure 1 is a summary diagram of the criteria.

The new criteria will be mandatory by being referred in the SOLAS and Load Line conventions for passenger and cargo ship of 24 m or more [2] and further improvements will be made by IMO in the future.

© Springer International Publishing AG, part of Springer Nature 2018
P. A. Wilson, *Basic Naval Architecture*,
https://doi.org/10.1007/978-3-319-72805-6_14

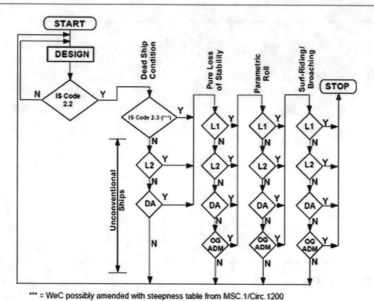

*** = WeC possibly amended with steepness table from MSC.1/Circ.1200

Fig. 1 Multi-tiered structure of the second generation intact stability criteria

2 The IMO Second Generation Intact Stability Criteria

It is a descriptive presentation of the criteria that is proposed here in order to give an idea of what type of physics are meant to be used for the criteria. It is likely that they receive further modifications before being agreed within the Correspondence Group.

2.1 Parametric Roll

This phenomenon refers to a severe righting arm variation under parametric conditions. The parameters are a wavelength to ship length ratio equal to 1, wave encounter frequency equal to twice the natural roll period and a disturbance to initiate the rolling motion.

2.1.1 First Level

At this stage the GM variation in waves is estimated through its values at lowest and highest draft corresponding respectively to wave crest and wave trough located at amidships. Figure 2 illustrates this simplification for wave crest amidships. If the variation of GM is higher than a standard value, R_{PR1}, then the vessel is considered as vulnerable.

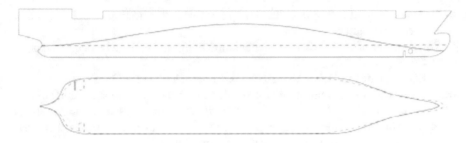

Fig. 2 Comparison of simplified waterline versus waterline of real wave in wave crest condition a

R_{PR1} represents the linear roll damping for a steady state or the linear roll damping and wave group effect combination for a transient state [1,3]. Its derivation is provided in [1]. Initially GM is considered to be the calm water GM for the the probabilityconsidered loading condition.

$$\frac{\Delta GM}{GM} \geq R_{PR1} \tag{1}$$

$$R_{PR} = 1.87, \quad \text{if } bilge\ radius < 0.01B \tag{2}$$

$$R_{PR} = 0.17 + 0.425 \left(\frac{100A_k}{LB}\right), \quad \text{if } C_m > 0.96 \tag{3}$$

$$R_{PR} = 0.17 + (10.625 \times C_m - 9.775)\left(\frac{100A_k}{LB}\right), \quad \text{if } 0.9 < C_m < 0.96 \tag{4}$$

$$R_{PR} = 0.17 + 0.2125 \left(\frac{100A_k}{LB}\right), \quad \text{if } C_m < 0.94 \tag{5}$$

L is the length between perpendiculars; B the moulded breadth; A_k the bilge keel area; C_m the midship area coefficient.

Variation of KB is neglected and only BM variations due to the difference of the waterline moments of inertia in crest and trough conditions at amidships are considered. Delta GM is not corrected for free surface effect [4] Annex 5 as opposed to initial GM which is corrected [5]. Therefore, the waterline moments of inertia in crest and trough conditions are approximated by the calm waterline moment of inertia at two different drafts:

$$\Delta GM = \frac{I_H - I_L}{2V}; \text{Only if, } \frac{V_D - V}{A_W(D - d)} \geq 1.0 \tag{6}$$

where I_H and I_L are the waterplane moment of inertia for higher and lower draughts, d_H and d_L calculated as per given in [6]; Δ is the volume of displacement for the considered loading condition; Δ_D is the volume of displacement at waterline equal to D; D is moulded depth; d is the draught at loading condition; A_W is the water plane area at draught d.

Wave steepness used to carry out calculation of d_H and d_L is to be decided among three existing proposals [4] Annex 15.

If Eq. (6) is not satisfied, the ΔGM is to be calculated as half the difference between maximum and minimum GM obtained with the wave crest centred at LCG and each $0.1L$ backward and forward up to $0.5L$. A method based on the American Bureau of Shipping guidelines [7] is provided in Annex 2. In this case the characteristics of the wave to be considered are very conservative:

$$\text{Wave length} : \lambda = L \tag{7}$$

$$\text{Wave height} : H = L \cdot S_W \text{ where, } S_W = 0.0167. \tag{8}$$

2.1.2   Second Level

The second level is made of two checks. The ship is required to comply with one of them.

First Check

The ship is considered vulnerable if the index C_1 is bigger than a defined standard R_{PR0} most likely taken as 0.06 [4] Annex 21 but still under discussion. C_1 is the weighted average from a set of waves and is defined for 16 wave cases as:

$$C_1 = \sum_{i=1}^{N} W_i C_i > R_{PR0} = 0.06 \tag{9}$$

where W_i is the weighting factor representing the probability of occurrence for each wave case and may be taken from Table 1.

Basically, the index C_i represents the physical vulnerability of the ship to parametric rolling for each wave case. Then, if this vulnerability is established, it is consistent to somehow reduce it accordingly to the probability of occurrence of such wave able to make the ship vulnerable.

C_i is taken as 1 if Eq. (10) or condition (11) is satisfied and 0 otherwise:

$$V_{PRi} \leq V_D \tag{10}$$

$$\frac{\Delta GM\,(H_i, \lambda_i)}{GM\,(H_i, \lambda_i)} \geq R_{PR1} \text{ and } GM\,(H_i, \lambda_i) \leq 0 \tag{11}$$

In Eq. (10), V_D is the design speed of the vessel and V_{PRi} is the ship speed for parametric resonance conditions, defined as per Eq. (12). 2:1 parametric resonance exists at this speed because the average natural roll period is exactly half the wave encounter frequency.

Table 1 Wave cases for parametric rolling evaluation

Wave case number	Weight W_i [m]	Wavelength λ_i (m)	Wave height H_i(m)
1	0.000013	22.574	0.350
2	0.001654	37.316	0.495
3	0.020912	55.743	0.857
4	0.092799	77.857	1.295
5	0.199218	103.655	1.732
6	0.248788	133.139	2.205
7	0.208699	166.309	2.697
8	0.128984	203.164	3.176
9	0.062446	243.705	3.625
10	0.024790	287.931	4.040
11	0.008367	335.843	4.421
12	0.002473	387.440	4.769
13	0.000658	442.723	5.097
14	0.000158	501.691	5.370
15	0.000034	564.345	5.621
16	0.000007	630.684	5.950

$$V_{PRi} = \left| \frac{2\lambda_i}{T_\phi} \cdot \sqrt{\frac{GM(H_i, \lambda_i)}{GM}} - \sqrt{g \frac{\lambda_i}{2\pi}} \right| \tag{12}$$

where T_ϕ is the roll natural period and GM is the calm water metacentric height.

In condition (11), $GM(H_i, \lambda_i)$ is the average GM value during the passage of a wave with length λ_i and height H_i and $\Delta GM(H_i, \lambda_i)$ represents the GM variation in a series of waves. Here, it is the true sinusoidal water line at different wave crest positions x_{C_j} and balanced trim and sinkage that is taken to work out the GM variation in waves, as illustrated by Fig. 3. It is the same procedure than the one used in case Eq. (6) is not satisfied at the first level. Again, the method based on the American Bureau of Shipping guidelines [7] provided in Annex 1 is suitable.

Second Check

The second check is based on response in head or following waves for the range of operational ship speeds and heading. The stability calculation is expected to account for heave and pitch quasi-statically. The ship is considered vulnerable if the index C_2 is bigger than $R_{PR1} = 0.15$.

$$C_2 = \left[\sum_{i=1}^{3} C_2(Fn_i, \beta_h) + C_2(0, \beta_h) + \sum_{i=1}^{3} C_2(Fn_i, \beta_f) \right] / 7$$

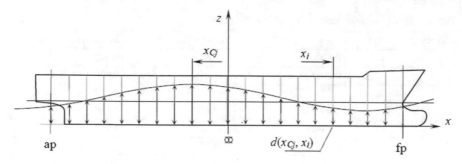

Fig. 3 Definition of the draft ith station with jth position of the wave crest

Table 2 Corresponding encounter speed factor K_i

i	K_i	Corresponds to encounter with
1	1.0	Head or following waves at V_s
2	0.866	Waves with 30° relative bearing to ship centreline at V_s
3	0.500	Waves with 60° relative bearing to ship centreline at V_s

$$C_2(Fn_i, \beta_i) = \sum_{i=1}^{N} W_i C_i$$

Subscripts h and f stands for head and following seas.

The weighting factor W_i is obtained for each wave case from a wave scatter diagram. The procedure is too cumbersome for being explained in this chapter. It will be explained in more details in part III.D.2. C_i is 0 if the maximum roll angle exceeds R_{PR2} (most likely 25°) and 1 if not. C_i is calculated in 6 different conditions made of three different speeds V_i and two wave relative headings: head waves corresponding to $C_2(Fn_i, \beta_h)$ and following seas corresponding to $C_2(Fn_i, \beta_f)$). The speeds V_i are calculated according to Eq. (13) and Table 2.

$$Fn_i = \frac{V_i}{\sqrt{Lg}}$$
$$V_i = Vs \cdot K_i \tag{13}$$

The wave conditions for the evaluation of the maximum roll angle (setting C_i) are as follows:

$$\text{Wave length} : \lambda = L \tag{14}$$

$$\text{Wave height} : h_j = 0.01 \cdot jL, \quad \text{where } j = 1, 2, \ldots, 10 \tag{15}$$

Three methods are proposed to determine the maximum roll angle depending on conditions mentioned below.

2.1.3 Numerical Transient Solution

This method should provide evaluation of the *natural response* and relies on obtaining solution of Mathieu-type equations (derived from mass–spring–damper system) either linear with single DOF (16) proposed in [4] Annex 15, or nonlinear with single DOF (17) proposed in [4] Annex 18. Solutions obtained through numerical simulation in the time domain using the conventional Runge–Kutta method.

$$\ddot{\Phi} + 2\delta_\Phi \cdot \dot{\Phi} + \omega^2 f_N(\Phi, t) = 0 \tag{16}$$

$$\ddot{\Phi} + 2\mu \cdot \dot{\Phi} + \beta \cdot \dot{\Phi}\left|\dot{\Phi}\right| + \delta \cdot \dot{\Phi}^3 + \omega^2 f_N(\Phi, t) = 0 \tag{17}$$

Φ is the roll angle; μ, β, δ are the linear, quadratic and cubic damping coefficients obtained from Ikeda's simplified method, ω is the undamped natural frequency and f_N is the restoring force calculated as so:

$$f_N(\Phi, t) = \frac{GZ_W(\Phi, t)}{GM}$$

$$GZ_W(\Phi, t) = GZ_{still}(\Phi) + \{GM(t) - GM_{still}\}\left\{\sin(\Phi) - \sin(\Phi)^3 / \sin(\Phi_{max})^2\right\}$$

$GZ_W(\Phi,t)$ is the righting lever calculated in waves with static balance in heave and pitch for instantaneous position of the wave crest at time t. Φ_{max} is the capsizing angle. This GZ calculation method is proposed in [4] Annex 17 as more accurate than other existing proposals (see [4] Annex 15 and [8] p. 110) but is not yet agreed because a more conservative estimation could be preferred at this level. In any case, the calculation is based on the Froude–Krylov assumption and does not take radiation with coupling and diffraction components into account.

2.1.4 Analytical Forced Response

This method aims to avoid the use of numerical simulation at Level 2 by analytically solving the equation of motion (18). The solution is of the form $\Phi(t) = A\sin(\omega t - \varepsilon)$ with ω being half the encounter frequency. The amplitude A, i.e. the steady states of principal parametric rolling, is found by solving the 12th order equation (19) for A with, e.g. Maki's method [9].

$$\ddot{\Phi} + 2\mu \cdot \dot{\Phi} + \delta \cdot \dot{\Phi}^3 + \omega_\Phi^2 \Phi + \omega_\Phi^2 l_3 \Phi^3 + \omega_\Phi l_5 \Phi^5 + \\ \omega_\Phi^2 \left(GM_{mean} + GM_{amp}\cos\omega_e t\right) \cdot \left(1 - \left(\frac{\Phi}{\pi}\right)^2\right)\frac{\Phi}{GM} = 0 \tag{18}$$

$$\left(\frac{\pi^2\omega\left(3A^2\omega^2\gamma + 8\alpha\right)}{\left(2\pi^2 - A^2\right)\omega_\Phi^2}\right)^2 + \left(\begin{array}{c}\frac{6A^2\omega_\Phi^2 - 8\pi^2\omega_\Phi^2}{4\left(\pi^2 - A^2\right)\omega_\Phi^2}\frac{GM_{mean}}{GM} + \\ \frac{-5\pi^2 A^4 l_5 \omega_\Phi^2 - 6\pi^2 A^2 l_3 \omega_\Phi^2 + 8\pi^2\omega^2 - 8\pi^2\omega_\Phi^2}{4(\pi^2 - A^2)\omega_\Phi^2}\end{array}\right)^2 = \left(\frac{GM_{amp}}{GM}\right)^2 \tag{19}$$

where γ and α are obtained following [4] Annex 15 appendix 1. Damping coefficients obtained from Ikeda's simplified method are required. [10] mentions the necessity to add additional component for lift of bilge keels and hull in forward speed. l_3 and l_5 are constant coefficients of third and fifth order restoring moment in calm water obtained from least fit square on GZ curve calm water. GM_{amp} and GM_{mean} are the amplitude of the variation of metacentre height and the mean of the variation, respectively. GM is the metacentre height in calm water.

This method is the one that is proposed in the explanatory notes [11] and that is likely to be kept for the regulatory application.

2.2   Pure Loss of Stability

This failure mode is related to the variation of roll restoring moment during a sufficiently long time to put ship into danger when travelling in waves [12].

The criteria are to be applied only to ships having a Froude number exceeding 0.24.

2.2.1   First Level

A ship is considered vulnerable if the minimum GM in waves is smaller than a specified standard R_{PLA}:

$$GM_{min} < R_{PLA} \tag{20}$$

where, R_{PLA} is to be defined but proposed as equal to 0.05 or $1.83\,dFn^2$.

The minimum GM is calculated as shown in Eq. (21):

$$GM_{min} = KB(d) + \frac{I_L}{V_d} - KG \tag{21}$$

where $KB(d)$ is the vertical position of the centre of buoyancy at draught d (loading condition under consideration) and I_L is the waterplane moment of inertia determined at draught d_L (wave crest condition) defined in Eq. (22):

$$
\begin{aligned}
d_L &= d - \delta d_L \\
\delta d_L &= Min\left(0.75d, \frac{L \cdot S_w}{2}\right)
\end{aligned}
\tag{22}
$$

where S_w is the wave steepness taken as equal to 0.0334 or from [4,6], Tables 5-B-1 or 5-B-2. It is generally twice the one used for Parametric Rolling.

The procedure is the same as for Level 1 Parametric Rolling regarding the solution of equation (22).

2.2.2   Second Level

A ship in a particular loading condition is considered as vulnerable if:

$$\max(CR_1, CR_2, CR_3) > R_{PL0} \tag{23}$$

$$CR_1 = \sum_{i=1}^{N} W_i C_{1_i} \tag{24}$$

$$CR_2 = \sum_{i=1}^{N} W_i C_{2_i} \tag{25}$$

$$CR_3 = \sum_{i=1}^{N} W_i C_{3_i} \tag{26}$$

and with W_i, a statistical weight taken from Table 3, [13].

The three criteria C_{1_i}, C_{2_i} and C_{3_i} are defined as follows:

If $GM_{\min} < R_{PL1}$, then C_{1_i} is taken as 1 and 0 otherwise. GM_{min} is calculated as explained above in Level 1 parametric roll when Eq. (22) is not satisfied.

Table 3 Environmental conditions for pure loss

Case number	Weight W_i	Wavelength λ_i [m]	Wave height H_i [m]	Wave steepness $s_{w,i}$	$1/s_{w,i}$
1	0.000013	22.574	0.700	0.0310	32.2
2	0.001654	37.316	0.990	0.0265	37.7
3	0.020912	55.743	1.715	0.0308	32.5
4	0.092799	77.857	2.589	0.0333	30.1
5	0.199218	103.655	3.464	0.0334	29.9
6	0.248788	133.139	4.410	0.0331	30.2
7	0.208699	166.309	5.393	0.0324	30.8
8	0.128984	203.164	6.351	0.0313	32.0
9	0.062446	243.705	7.250	0.0297	33.6
10	0.024790	287.931	8.080	0.0281	35.6
11	0.008367	335.843	8.841	0.0263	38.0
12	0.002473	387.440	9.539	0.0246	40.6
13	0.000658	442.723	10.194	0.0230	43.4
14	0.000158	501.691	10.739	0.0214	46.7
15	0.000034	564.345	11.241	0.0199	50.2
16	0.000007	630.684	11.900	0.0189	53.0

If $\Phi_{loll} > R_{PL2}$, then C_{2_i} is taken as 1 and 0 otherwise. Φ_{loll} is a maximum loll angle determined from the GZ calculated with the wave crest centred at the LCG, and each $0.1L$ interval forward and backward of the LCG up to $0.5L$.

If $GZ_{\max} < R_{PL3}$, then C_{3_i} is taken as 1 and 0 otherwise. GZ_{max} is the minimum value of the maximum righting arm determined from GZ curves obtained as mentioned in C_2 definition.

Several proposals exist for the value or the standards.

The latest agreed ones available from [13,14] are:

$$R_{PL0} = 0.05 \tag{27}$$

$$R_{PL1} = 0.06\,\text{m} \tag{28}$$

$$R_{PL2} = 30° \tag{29}$$

$$R_{PL3} = 8\left(\frac{H_i}{\lambda_i}\right) dF_n^2 [\text{m}] \tag{30}$$

2.3 Dead Ship Stability

This mode is considered with the ship without propulsive power submitted to high seas and strong winds. These two exciting forces may cause synchronous roll resonance and large heeling moments and possibly lead to stability failure.

2.3.1 First Level

The criterion is agreed as the current weather criterion of the 2008 Intact Stability Code with the modified Table 3 of wave steepness included in MSC.1/Circ. 1200.

A ship is considered as not vulnerable if area b is equal or greater than area a as indicated on Fig. 4.

I_{W1} is the steady wind heeling lever and I_{W2} is the gust wind heeling lever. Φ_0 is the equilibrium angle of heel, Φ_1 is the angle of roll windward owing to wave (See [15] p. 15 for its calculation), Φ_2 is the angle of down flooding or second intercept with gust heeling moment or $50°$.

The wind heeling moment is calculated according to (31):

$$
\begin{aligned}
I_{W1} &= \frac{P \cdot A \cdot Z}{1000 \cdot g \cdot \Delta} \\
I_{W2} &= 1.5 \cdot I_{W1}
\end{aligned}
\tag{31}
$$

P is the wind pressure (504 Pa), A is the projected lateral area of ship and cargo above the waterline (m^2), Z is the vertical distance between A and the surface of underwater lateral area (m), Δ is the displacement (t).

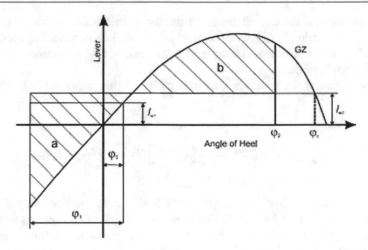

Fig. 4 Vulnerable ship stability curve

2.3.2   Second Level

Many discussions are still ongoing for the second level criterion of this failure mode. Their last versions are contained in the following references: [8] Annex 3 (Method A) and 4 (Method B) and [16] Annex 7 and [17] Annex 19 and 20.

However, what seems to be agreed is described below.

The roll motion of the ship is described by the following 1 DOF nonlinear model:

$$(J_{xx} + J_{add}) \ddot{\Phi} + D(\dot{\Phi}) + \Delta \cdot GZ(\Phi) = M_{wind,tot}(\Phi, t) + M_{waves}(t) \qquad (32)$$

J_{xx} is the ship dry moment of inertia, J_{add} is the added moment of inertia, $D(\dot{\Phi})$ is the general damping moment, Δ is the ship displacement, GZ (Φ) is the restoring lever, $M_{wind,tot}(\Phi, t)$ is the total instantaneous moment due to wind taking into account also the effect of the hydrodynamic reaction, $M_{waves}(t)$ is the total instantaneous moment due to waves.

The following procedure takes into account the consolidation of two existing methods [2]. Equation (32) is then re-arranged as below:

$$\ddot{\Phi} + 2\mu \cdot \dot{\Phi} + \beta \cdot \dot{\Phi} |\dot{\Phi}| + \delta \cdot \dot{\Phi}^3 + \omega_0^2 c(\Phi) = \omega_0^2 \cdot \left(\bar{m}_{wind,tot} + m(t) \right) \qquad (33)$$

With forcing $m(t) = \delta m_{wind,tot}(t) + m_{waves}(t)$ and restoring $c(\Phi) = {GZ(\Phi)}/{GM}$ $\bar{m}_{wind,tot}$ is mean wind heeling lever and $\delta m_{wind,tot}(t)$ is the gust wind heeling lever.

$$\bar{m}_{wind,tot} = 0.5 \cdot \frac{\rho_{air} \cdot \bar{V}_W^2 \cdot (H_W + H_{yd}) \cdot C_y \cdot A_L}{\Delta \cdot GM} \qquad (34)$$

where ρ is the air volumetric mass density, V_w is the mean wind speed, H_W is the vertical distance from the waterline to the assumed centre of wind pressure (positive above water), H_{yd} is the vertical distance from the waterline to the assumed centre of the sway drift reaction (positive below water), C_y is the lateral wind drag coefficient, A_L is the lateral windage area, Δ is the ship displacement, GM is the ship metacentric height in upright position.

The spectrum of the total non-dimensional fluctuating moment is calculated as follows:

$$S_{m_{waves}}(\omega) = r^2(\omega) \cdot \frac{\omega^4}{g^2} \cdot S_{ZZ}(\omega)$$

$$S_{\delta m_{wind,tot}}(\omega) = \left(\frac{\rho_{air} \cdot \bar{V}_W^2 \cdot (H_W + H_{yd}) \cdot C_y \cdot A_L}{\Delta \cdot GM} \right)^2 \cdot \chi^2(\omega) \cdot S_v(\omega) \qquad (35)$$

$$S_m(\omega) = S_{m_{waves}}(\omega) + S_{\delta m_{wind,tot}}(\omega)$$

where $r(\omega)$ is the effective wave slope function, ω is the wave frequency (Doppler effect due to drift is neglected), $S_{zz}(\omega)$ is the sea elevation spectrum, $\chi(\omega)$ is the aerodynamic admittance function, $S_v(\omega)$ is the wind gustiness spectrum [3] Annex 3. The spectra to be used are still to be defined.

Under some proper assumptions, the equation of motion can be solved. Details may be found in [18, 19].

Once the spectrum of the roll motion is obtained, a time-dependent capsize index will be determined. This weighted average total stability failure probability is to be compared to a standard R_{DS} equal to 10^{-6} as recommended by [20] p. 15 but to be confirmed as not yet agreed within the CG.

2.4 Excessive Acceleration

The criteria here may be expressed in terms of maximum allowable GM or maximum allowable lateral acceleration value at a specified point on the ship. For some ships, satisfying such conditions might be quite difficult because damage stability requirements may be higher than the GM values calculated here.

2.4.1 First Level

When calculating the variance of the roll angle in a natural seaway for a specified wave spectra, it can be seen that the dominating contribution to one of the integral involved in the calculation (Eq. (24) in [17]), comes from the region of frequencies close to the natural roll frequency. Therefore, the variance of roll is simplified to the following expression:

$$\sigma_\varphi^2 = \frac{\omega_\varphi^4 I_1}{g^2} \cdot \pi^2 r^2 \omega_\varphi S_\zeta \frac{(\omega_\varphi)}{\delta_\varphi} = 0.0256 r^2 \omega_\varphi^5 S_\zeta \frac{(\omega_\varphi)}{\delta_\varphi} \qquad (36)$$

where ω_φ is the natural roll frequency; I_1 an integral considering the influence of the main wave direction and directional spreading of wave energy; r the reduction factor of the exciting heel moment; S_ς spectrum; δ_φ the logarithmic decrement of free roll decay at the amplitude φ.

Finally leading to the following equations for lateral acceleration and lateral jolt: $\sigma_a = \sigma_\varphi(g + \omega_\varphi^2 h)$ and, $\sigma_j = \sigma_\varphi \omega_\varphi(g + \omega_\varphi^2 h)$.

Two standards have been proposed. They are 5.9 (m/s) or 7.848 (m/s) for ships with the length between perpendiculars greater than 250.0 m.

2.4.2   Second Level

The roll equation in linearized form reads as follow:

$$I_\Phi \ddot{\varphi} + b_\varphi \dot{\varphi} + c_\varphi \varphi = M \sin(\omega_e t + \varepsilon) \tag{37}$$

where I_φ is the roll moment of inertia (including dry and added mass), b_φ is the effective roll damping coefficient depending on roll amplitude φ_a, c_φ is the effective linearized stiffness, M is the amplitude of exciting moment. The obtaining of the coefficients and amplitude of forcing together with the considered assumptions is precisely described in [17] p. 20. The solution of the equation under these assumptions can be shown to be:

$$\frac{\varphi_a}{\zeta_a} = \frac{r\omega^2 \omega_\varphi^2 \sin \mu}{g \left[(\omega_\varphi^2 - \omega^2)^2 + \omega^2 \omega_\varphi^2 \left(\delta_\varphi / \pi \right) \right]^{1/2}} \tag{38}$$

where ζ_A is the wave amplitude, r the reduction factor of the exciting heel moment, ω_φ the natural roll frequency, ω the wave frequency, μ the wave direction, g the gravitational acceleration and δ_φ the logarithmic decrement of free roll decay at the amplitude φ.

It follows that lateral acceleration amplitude is: $\frac{a_y}{g} = k_L \varphi_a \left(1 + h\omega^2/g \right)$, and that lateral jolt amplitude is: $\frac{J_a}{g} = k_L \varphi_a \omega \left(1 + h\omega^2/g \right)$.

where k_L is a factor depending on longitudinal position on the ship and h is the height of the observation point. The use of the wave spectrum to determine the variance of the roll angle in natural seaways would not be developed here.

Concerning the standards, reference was firstly made to existing standards of the High Speed Craft Code but it was then mentioned as inappropriate [4] Annex 29. The value of 0.2 g could be read.

Environmental condition proposals are contained in appendixes 1, 2 and 3 of [4] and appendix 1 of [21].

3   Direct Stability Assessment (DSA)

A summary of the main requirements expressed by the Working Group for the DSA is proposed below. According to those, [22] proposed a six DOF model that should suit all failure modes.

The Correspondence Group is currently developing and agreeing on the DSA procedure. They are contained in [17] Annex 21 and [23]. The computational tool must consist of time-domain simulation software with the following minimum capabilities.

3.1   General Requirements

1.a.i. Waves modelling may be based on small waves theory (airy waves) unless mature nonlinear wave model are available. Statistic and hydrodynamic validity of wave model must be recognized if Fourier series are used for mathematical duplication. (cf. SLF 51/INF 4).

  ii. Roll damping must include wave, vortex and skin friction component. Roll decay test should be preferred, CFD otherwise. Special care is to be given to duplication: wave component of damping already included in calculation of diffraction forces. Or with component of damping (e.g. cross-flow drag) directly computed that must be excluded from calibration data used.

  iii. Hull forces are resistance forces (including wave, vortex and skin friction) and sway and yaw forces and moments. If the radiation and diffraction forces are calculated as a solution of the hull boundary-value problem, measures must be taken to avoid including these effects more than once.

3.2   Parametric Roll and Excessive Acceleration

1.a.i. At least 3 DOF (heave, roll and pitch) and 5 DOF for some ships (adding sway and yaw).

  ii. Froude–Krylov and hydrostatic forces should be calculated with body exact formulations (panel and strip-theory approach).

  iii. Radiation and diffraction forces should be calculated with approximate coefficients, body linear formulation or body exact solution of the boundary-value problem.

  iv. Careful not to account damping component more than once (see general requirements).

  v. Capability to reproduce statistically valid long-crested irregular waves must be present.

3.3 Pure Loss of Stability

1.a.i. At least 4 DOF (surge, heave, roll and pitch) modelled and full coupling for the non-modelled ones.
  ii. Froude–Krylov and hydrostatic forces should be calculated from body exact formulations including panel and strip-theory.
  iii. Radiation and diffraction calculated as for PR and excessive Acceleration. Hydrodynamic forces due to shedding vortex from hull are to be properly modelled.
  iv. Thrust may be obtained from coefficient-based model (accounting for propeller hull interaction).
  v. Resistance effects must not be included more than once.

3.4 Surf-Riding and Broaching-to

1.a.i. At least 4 DOF (surge, sway, roll and yaw) required and full coupling or static equilibrium assumed for other DOF.
  ii. Froude–Krylov and hydrostatic forces must be calculated from body exact formulations including panel and strip-theory or polynomial equations.
  iii. Radiation and diffraction calculated as for all other modes.
  iv. Hydrodynamic forces due to shedding vortex from a hull should be properly modelled. This should include hydrodynamic lift forces and moments due to the coexistence of wave particle velocity and ship forward velocity, other than manoeuvring forces and moments in calm water.
  v. Resistance, roll damping, sway and yaw forces and moments should not be included more than once.

3.5 Dead Ship Condition

1.a.i. At least 5 DOF (i.e. all except surge).
  ii. Froude–Krylov and hydrostatic forces are to be calculated from body exact formulations including panel and strip-theory.
  iii. Radiation and diffraction calculated as for all other modes.
  iv. Three component aerodynamic forces and moments should be evaluated based on model test results unless CFD shows sufficient agreement.
  v. Longitudinal drift force, drift heeling moment and drift yawing moment obtained from experiments unless CFD shows sufficient agreement.
  vi. Optional capability to reproduce statistically valid long-crested or short-crested irregular waves.
  vii. Roll damping, sway and yaw forces and moments should not be included more than once.

References

1. V. Belenky, C.C. Bassler, K.J. Spyrou, *Development of Second Generation Intact Stability. Criteria, Naval Surface Warfare Center, US Navy*
2. Proceedings of The 11th International Ship Stability Workshop, Current Status of New Generation Intact Stability Criteria Development, Alberto Francescutto, University of Trieste, Naoya Umeda Osaka University
3. Current Status of Second Generation Intact Stability Criteria Development and Some Recent Efforts, Naoya Umeda Osaka University
4. IMO SLF55/INF15
5. Revision Proposals of Draft Vulnerability Criteria and Standards For Parametric Roll and Pure Loss of Stability Japan Submission to the Iscg, 3rd March 2014
6. Proposed Amendments to Part B of the 2008 IS Code to Assess The Vulnerability of Ships to the Parametric Rolling Stability Failure Mode, IMO, 2014
7. Guide for the Assessment of Parametric Roll Resonance in the Design of Container Carriers American Bureau Of Shipping, June 2008
8. IMO SLF52/INF2
9. A. Maki, N. Umeda, S. Shiotani, E. Kobayashi, Parametric rolling prediction in irregular seas using combination of deterministic ship dynamics and probabilistic wave theory. J. Mar. Sci. Technol. **16**(13), 294310 (2011)
10. S. Kruger, H. Hatecke, H. Billerbeck, A. Bruns, F. Kluwe, Investigation of the 2nd generation of intact stability criteria for ships vulnerable to parametric rolling in following seas. Hamburg University of Technology Flensburger Schiffbau-Gesellschaft(2013). Accessed 9–14 June 2013
11. Draft Explanatory Notes on the Vulnerability of Ships to the Parametric Rolling Stability Failure Mode, IMO 2014
12. V.N. Belenky, N.B. Sevastianov, *Stability and Safety of Ships Risk of Capsizing*, 2nd edn. (Jersey City, SNAME, 2007), pp. 195–198
13. H. Sadat-Hosseini, F. Stern, A. Olivieri, E. Campana, H. Hashimoto, N. Umeda, G. Bulian, A. Francescutto, Head-waves parametric rolling of surface combatant. Ocean Eng. **37**, 859–878 (2010)
14. Proposed Amendments to Part B of the 2008 IS Code to Assess the Vulnerability of Ships to the Pure Loss of Stability Failure Mode, IMO 2014
15. IMO MSC267.85
16. IMO SLF53/INF10
17. IMO SLF54/INF12
18. G. Bulian, A. Francescutto, A simplified modular approach for the prediction of the roll motion due to the combined action of wind and waves. J. Eng. Marit. Environ. **218**(M3), 189–21 (2004)
19. G. Bulian, A. Francescutto, A. Maccari, Possible simplified mathematical models for roll motion in the development of performance-based intact stability criteria extended and revised version. Quaderno Di Dipartimento N 46, Department Dinma, University Of Trieste (2008)
20. IMO SLF54/Wp3
21. IMO SLF55/WP3
22. G. Bulian, A. Fancescutto, Second generation intact stability criteria: on the validation of codes for direct stability assessment in the framework of an example application. Pol. Marit. Res
23. Draft Guidelines of Direct Stability Assessment Procedures as a Part of the Second Generation Intact Stability Criteria, SDC 1. INF8. Annex 27, IMO, 2014

Examples

15

1 Examples 1

1. A oil tanker has a moulded beam of 39.5 m with a moulded draft of 12.75 m and a midship area of 496 m^2. Calculate the midship area coefficient C_m.
   [$C_m = 0.9849$]
2. Find the area of the waterplane of a ship that is 36 m long, 6 m beam that has a fineness coefficient of 0.8.
   [172.8 m^2]
3. The following data in Table 1 relates to ships the late Victorian era. The units are in *feet*.
   Calculate for each ship type ∇, A_m, A_w in *SI* units. You may use 1 ft $= 0.3048$ m.
4. A ship is 150 m long, with a beam of 20 m and load draft of 8 m, light draft 3 m. The block coefficient is 0.788 for load draft and 0.668 for light draft. Calculate the two different displacements.
   [18912 m^3, 6012 m^3]
5. A ship 100 m long, 15 m beam and a depth of 12 m is floating at even keel with a draft of 6 m with block coefficient of 0.800 in standard salt water of density 1.025 t.m^{-3}. Find out how much cargo has to be discharged if the ship is to float at the same draft in freshwater.
   [180t]
6. A ship of 120 m length, with a 15 m beam has a block coefficient of 0.700 and is floating at the load draft of 7 m in freshwater. How much extra cargo can be loaded if the ship is to float at the *same* draft but in standard density sea water 1.025 t.m^{-3}
   [In salt water 9040.5 t and freshwater 8820 t]
7. A general cargo vessel with the following particulars; length between perpendiculars, 120 m, midship breadth 20 m, draft 8 m, displacement Δ 14, 000 t, midship

© Springer International Publishing AG, part of Springer Nature 2018
P. A. Wilson, *Basic Naval Architecture*,
https://doi.org/10.1007/978-3-319-72805-6_15

Table 1 Ship information

Class of ship	Length	Breadth	Draught	C_b	C_w	C_m
Pacific S. N. Co.	390.0	42.5	21.0	0.609	0.77	0.86
Royal mail Co.	344.0	40.5	21.0	0.590	0.76	0.84
National line Co.	385.0	42.0	22.0	0.659	0.80	0.88
Anchor line	350.0	35.0	21.0	0.687	0.84	0.85
Battle ship (3 decks)	260.0	61.0	25.5	0.537	0.82	0.76
Battle ship (2 decks)	238.0	55.8	23.8	0.530	0.84	0.66
Wood frigate	251.0	52.0	21.3	0.453	0.79	0.64
Sloop	160.0	31.3	13.0	0.495	0.76	0.79
Iron clad frigate	337.3	50.3	23.0	0.490	0.73	0.79

area coefficient of 0.985 and waterplane area coefficient 0.808 is to lengthened by 10 m in the midship position. Calculate the new values for C_b, C_w, C_p and Δ.
[$C_b = 0.733$, $C_w = 0.823$, $C_p = 0.744$, $\Delta = 15620\,\text{t}$]

2 Examples 2

1. A ship has a displacement volume $\nabla = 15500\,\text{m}^3$, $C_B - 0.71$ and $C_M = 0.95$. The area of the immersed midship section is $A_M = 160\,\text{m}^2$, and the ratio of beam to draught is 2.17. Determine the length, beam and draught of the ship.
   [$L = 129.6\,\text{m}$, $B = 19.12\,\text{m}$, $T = 8.81\,\text{m}$]
2. A vessel has a length of 130 m, beam of 17.7 m and a draught of 7.3 m. The displacement volume is $\nabla = 12044.5\,\text{m}^3$ and $C_P = 0.845$. Determine the area of the immersed midship section A_M and C_M.
   [$A_M = 109.6\,\text{m}^2$, $C_M = 0.849$]
3. A two-man sailing dinghy displaces a volume of $0.27\,\text{m}^3$. Its load waterline length is 4.25 m, and the waterline beam at amidships is 1.1 m. Assuming $C_P = 0.58$ and a midship sectional area coefficient $C_M = 0.63$ calculate values of $\textcircled{m}$, C_B and draught (centreboard raised).
   [$\textcircled{m} = 6.576$, $C_B = 0.365$, $T = 0.158\,\text{m}$]
4. A luxury motor yacht is to have a displacement volume of $350\,\text{m}^3$ and a beam to draught ratio of 3.5. Assuming a Froude number $F_n = 0.40$ use the design trend lines to estimate suitable length (LBP), beam (B) and draught (T) for this vessel. Also calculate the vessel speed corresponding to the proposed Froude number.
   [approx. $LBP = 46.86\,\text{m}$, $B = 7.7\,\text{m}$, $T = 2.2\,\text{m}$, $V = 16.67$ knots]
5. A guided missile destroyer is to have a displacement of 5000 tonnes and a maximum speed of 32 knots. A suitable beam to draught ratio is 3.0. Assuming a saltwater density of $1025\,\text{kg.m}^{-3}$ calculate suitable principal dimensions for this vessel. Note that you will need to make use of the design trend lines; however,

Table 2 Ship components weights

ITEM	Mass (kg)	$VCG(m)$	$LCG(m)$
Hull structure	720	1.0	0.3
Ballast keel	1100	−1.8	0.7
Aux.Eng. and Stern gear	150	0.1	−2.3
Deck gear, Anchor, Cables	100	1.6	2.0
Sails, mast, rigging	80	6.5	0.7
Crew	400	1.3	−3.0
Fuel, water, stores	280	0.2	−0.8

as no value is given for F_n, you will have to assume a typical value for $\bar{m}$ or C_B, say from Table 1.2. [*Hint: to get to the correct answer use an iterative technique, such as Newtons' method*]

[approx. $LBP = 125.5$ m, $B = 17.2$ m, $T = 5.7$ m]

6. The data given below in Table 2 relates to a 10 m sailing yacht. Use this information to calculate the position of the Centre of Gravity of this boat, both vertically and longitudinally, in the specified load condition. Also calculate the vertical distance through which the Centre of Gravity moves as a crew member of 80 kg mass is lifted to the mast head 13.5 m above the waterline in order to repair a rigging fault.

    VCG is measured from the load waterline, positive upwards. LCG is measured from amidships, positive forwards

    [$VCG = 0.004$ m, $LCG = -0.19$ m, Change of $VCG = 0.34$ m]

3   Examples 3

1. A fully ballasted ship model is floating freely in a towing tank. In transferring a ballast weight to change the trim of the model the weight is accidentally dropped into the tank. Does the water level in the tank: (a) rise, (b) fall or (c) stay the same? Give the reason for your answer.

2. The volume of water required to raise the water level in an empty lock from the downstream level to the upstream level is V_0. How much water is lost downstream in one locking cycle?:

    a. If one vessel of displaced volume V_1 passes upstream through the lock?
    b. If one vessel of displaced volume V_1 passes downstream through the lock?
    c. If one vessel of displaced volume V_1 passes upstream and a second of the same volume passes downstream through the lock?

Show details of how answers to (a), (b) and (c) were obtained.

The water losses on a canal system carrying barge traffic in both directions were monitored over a long period of time, and it was found that the average water loss is less than V_0 per lock cycle. Explain why this may be so.

3. The length, beam and mean draught of a ship are 115, 15.65 and 7.15 m, respectively. The midship sectional area and block coefficients are 0.921 and 0.665, respectively. The ship floats in salt water. Obtain:

   a. the displacement (mass) in tonnes,
   b. the displacement (weight) in MN,
   c. the midship sectional area,
   d. the prismatic coefficient.
   [$\Delta = 8771.3$ tonnes $= 86.05MN$, $A_M = 103.1\,\mathrm{m}^2$, $C_P = 0.722$]

4. An experiment on tethered mine motions is to be conducted in freshwater in a wave tank. The mine is a sphere of 5 cm diameter made of polystyrene (weight assumed negligible). The anchoring block is a lead cube of side length 5 cm. The specific gravity of lead is 14.0. If the mine is joined to the lead cube by a light cord such that it floats completely submerged just below the free surface, determine:

   a. the tension in the cord
   b. the reaction force between the lead cube and the bottom of the tank.
   [(a) $0.642N$, (b) $15.3N$]

5. A uniform rectangular block of density ρ floats completely immersed with two of its faces horizontal in two liquids which do not mix. The densities of the liquids are ρ_1 and ρ_2, such that $\rho_1 < \rho < \rho_2$. Show that the ratio in which the vertical faces of this block are divided by the common surface of the liquids is:

$$\frac{\rho_2 - \rho}{\rho - \rho_1}$$

6. A yacht moves from sea water to freshwater and sinks 0.5 cm. A man weighing $750N$ gets out of this yacht, and it rises to its original waterline. Obtain:

   a. the original weight of the yacht, including the man
   b. the waterplane area.
   [(a) $30750N$, (b) $15.3\,\mathrm{m}^2$]

4 Examples 4

[Where appropriate answers are given in square brackets]

1. A vessel has a displacement of 2643 tonnes. $TPM = 720$ relative to water of density $1025\,kg.m^{-3}$. The mean draught of this vessel, in water of density $1025\,kg.m^{-3}$, is 3.98 m. Obtain the mean draught of this vessel if it were launched in water of density $1006\,kg.m^{-3}$.
   [4.05 m]

2. For a wall-sided vessel of weight W the Centre of Gravity G is at a distance h above the centre of buoyancy B. When cargo of weight w is placed on board at a depth b below the original waterline, the draught increases by an amount of d. Show that the vertical distance between centres of gravity and buoyancy, after the addition of cargo w, is reduced by:

$$\frac{w}{W+w}(h+b+0.5d)$$

3. For an isosceles triangle, see Fig. 1, with a vertex at the origin as shown, establish the following second moments of area:
   $J_T = \frac{1}{48}B^3H$ about $y = 0$.
   $J_{OL} = \frac{1}{4}BH^3$ about $x = 0$.
   $J_L = \frac{1}{36}BH^3$ about axis through centroid (i.e. $\bar{x} = 2H/3$)
   $J_{1L} = \frac{1}{12}BH^3$ about base $x = H$.

4. A lamina is formed between $x = -\frac{L}{2}$ and $x = \frac{L}{2}$ bounded by:

   - the x-axis and,
   - the curve shown $y = B\cos(\frac{\pi x}{L})$

   in Fig. 2,

Fig. 1 Examples 3 question 1

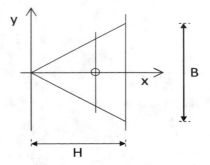

Fig. 2 Examples 3
question 4

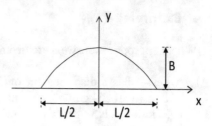

Show that the area of the lamina is $A = \frac{2}{\pi}LB$ and that the coordinates of the
centroid are:

$$\frac{\bar{x}}{B} = 0, \qquad\qquad \frac{\bar{y}}{B} = \frac{\pi}{8}$$

Note: $\cos 2\theta = 1 - 2\sin^2\theta$

5 Examples 5

[Where appropriate answers are given in square brackets]

1. A curve of waterplane areas for a ship can be approximately represented as shown
   in Fig. 3:

   $$A(z) = az^n$$

   where a is a constant. Show that in terms of the waterplane area coefficient C_W
   and block coefficient C_B:

   a.

   $$n + 1 = \frac{C_W}{C_B}$$

   b.

   $$\frac{KB}{T} = \frac{\bar{z}}{T} = \frac{C_W}{C_W + C_B}$$

Fig. 3 Examples 5
question 1

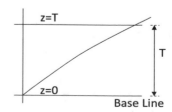

Fig. 4 Examples 3
question 2

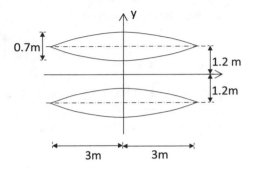

0.7m

1.2 m

1.2m

3m   3m

Fig. 5 Examples 3
question 6

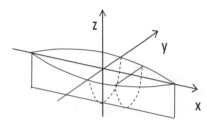

2. A catamaran has a waterplane as shown in Fig. 4,
   The waterplane area coefficient $C_W = 0.75$ for each demi-hull. Calculate the second moments of waterplane area J_L and J_T, about $x = 0$ and $y = 0$, respectively, assuming the waterplane is symmetric about $x = 0$ and each half waterplane is of the form.

$$y = \pm \frac{b}{2}\left[1 - \left(\frac{2x}{L}\right)^n\right]$$

   (y measured from the demi-hull Centre Line)
3. A Wigley mathematical hull form has parabolic waterlines and sections and is shown in Fig. 5. It is defined mathematically by the equation:

$$y = \frac{B}{2}\left[1 - \left(\frac{z}{T}\right)^2\right]\left[1 - \left(\frac{2x}{L}\right)^2\right].$$

   over the ranges $0 \geq z \geq -T, -s\frac{L}{2} \leq x \leq \frac{L}{2}$ and where B is the load WL breadth amidships and T is the draught.

   Show that the sectional curve area is

$$a(x) = \frac{2}{3}BT\left[1 - \left(\frac{2x}{L}\right)^2\right]$$

   (for $x =$ constant for any given section).

that the waterplane area is:

$$A(z) = \frac{2}{3}LB\left[1 - \left(\frac{z}{T}\right)^2\right]$$

(for z = constant for any given waterplane).

Hence, find values of the coefficients C_W, C_M, C_B, C_P and the depth of the centre of buoyancy below the load waterline.

$$\left[C_W = C_M = C_P = \frac{2}{3}, C_B = \frac{4}{9}, \frac{\bar{z}}{T} = -\frac{3}{8}\right]$$

---

6 Examples 6

Each numerical integration **must** be shown in tabular form:

1. Use Simpson's **First** rule to estimate the definite integral with THREE points.

$$\int_0^2 x^2 dx$$

Compare your answer with the exact value from integration.

2. Use Simpson's **First rule** to estimate the definite integral with SEVEN points.

$$\int_0^2 x^2 dx$$

Compare your answer with the exact value from integration.

3. Use Simpson's **First** rule to estimate the definite integral with NINE points.

$$\int_0^2 x^6 dx$$

Compare your answer with the exact value from integration.

4. Use Simpson's **First** rule to estimate the definite integral with FIFTEEN points.

$$\int_0^2 x^6 dx$$

Compare your answer with the exact value from integration.

5. Use Simpson's **Second** rule to estimate the definite integral with FOUR points.

$$\int_0^2 x^2 dx$$

Compare your answer with the exact value from integration.

6. Use Simpson's **Second** rule to estimate the definite integral with SEVEN points.

$$\int_0^2 x^2 dx$$

Compare your answer with the exact value from integration and the answer from question 2.

7. Use Simpson's **First** rule to estimate the definite integral with SEVEN points.

$$\int_0^2 \sin(5x) dx$$

Compare your answer with the exact value from integration.

8. Use Simpson's First rule to estimate the definite integral with THIRTEEN points.

$$\int_0^2 \sin(5x) dx$$

Compare your answer with the exact value from integration.

9. Use Simpson's **First** rule to evaluate:

$$\int_1^3 y dx$$

where the values of y and x are tabulated below:

x	1.00	1.25	1.50	1.75	2.00	2.25	2.50	2.75	3.0
y	2.45	2.80	3.44	4.20	4.33	3.97	3.12	2.38	1.80

7 Examples 7

[Where appropriate answers are given in square brackets]

1. The half-breadths (HB) at the loaded waterline of a vessel LBP = 144 m are as follows:

Station	0(AP)	1	2	3	4	5	6	7	8(FP)
HB(m)	17.0	20.8	22.4	22.6	21.6	18.6	12.8	5.6	0.0

Obtain the area and centroid of the loaded waterline as well as the longitudinal
and transverse second moments of area J_L and J_T.
[$4812\,m^2$, $13.68\,m$ abaft station 4, $5587765\,m^4$ about LCF, $641412\,m^4$ about
CL]
An appendage is added abaft AP with the following properties at the loaded
waterline: Length $= 21.6\,m$
Area $= 474\,m^2$
Centroid $= 8.17\,m$ abaft AP
$j_L = 44006\,m^4$ about AP
$j_T = 28440\,m^4$ about CL.
Obtain the area and centroid of the loaded waterline as well as the longitudinal
and transverse second moments of area J_L and J_T, including the appendage.
[$5286\,m^2$, $19.64\,m$ abaft station 4, $7508200\,m^4$ about new LCF, $669852\,m^4$
about CL]

2. The ordinates of the sectional area (a) curve for a vessel $LBP = 150\,m$ are given
as a percentage of the amidships sectional area as follows:

Station	0(AP)	$\frac{1}{2}$	1	2	3	4	5
$a(\%)$	1.70	15.5	33.2	65.6	88.1	97.8	100.0

Station	6	7	8	9	$9\frac{1}{2}$	10(FP)
$a(\%)$	100.0	98.0	83.3	42.0	17.3	0.0

The beam of the ship is 20 m, the draught 8 m and the midship sectional area
coefficient 0.96. Calculate the displacement volume and LCB.
[$16418.7\,m^3$, $2.29\,m$ forward of amidships]

3. A fuel tank is 6 m long and of uniform cross section. The cross section is defined
by the following breadths (B) at waterlines separated by 0.5 m (i.e. 0.5 m between
waterlines 0 and 1, 1 and 2):

Waterline	0	$\frac{1}{2}$	1	2	3	4	5
B(mm)	610	970	1310	1760	2130	2470	2740

Calculate the mass of oil of specific gravity 0.88 which can be carried with the
tank full and its vertical Centre of Gravity.
[24.75 tonnes, 1.48 m above WL 0]

8 Examples 8

[Where appropriate answers are given in square brackets]

1. A 110 m long (LBP) ship floats at an even keel draught of 4.85 m with the LCF at 1.30 m forward of amidships. A mass of 30 tonnes is moved forward until the draught at FP is 5.05 m. Obtain the final position of this mass, given that its initial position was 47.0 m abaft amidships and MCT is 64 tonne.m/cm.
   [40.4 m forward of amidships]
2. A 140 m long (LBP) ship floats with draughts of 4.85 m aft and 4.25 m forward. LCF is 1.90 m forward of amidships. TPC and MCT are 19.4 and 108 tonne.m/cm, respectively. Find where a mass of 135 tonnes should be placed in order to maintain the draught at FP at 4.25 m. Obtain the new draught at AP.
   [9.55 m aft of amidships; 4.993 m]
3. A 120 m long (LBP) ship, with maximum beam of 15 m, is floating in standard salt water at draughts of 6.6 m forward (FP) and 7.0 m aft (AP). The block and waterplane area coefficients are both 0.75. LCF is assumed at amidships, and GM_L is 120 m. Find how much more cargo can be loaded and its position relative to amidships, if the ship has to cross a bar with maximum draught of 7.0 m.
   [276.75 tonnes, 13.6 m forward of amidships]
4. A large motor yacht, prepared for an extended charter, has the following properties in Table 3:
   $LBP = 40$ m.
   $TPC = 2.09$.
   $MCT = 4.37$ tonne.m/cm
   $LCF = 3.5$ m aft of amidships.
   Initial draught aft $T_A = 1.80$ m measured 1.0 abaft AP.
   Initial draught forward $T_F = 1.68$ m measured at FP.
   Calculate the forward and aft draughts once the following items are put on board:
   [$T_A = 1.96$ m, $T_F = 2.05$ m]
5. A new class of frigate has the following hydrostatic particulars at a mean draught of 4 m relative to water of density 1025 kg.m^{-3}:
   $TPM = 720$
   $MCT = 3000$ tonne.m/m
   $LCF = 4$ m abaft amidships.

Table 3 Ship particulars

ITEM	Mass (tonnes)	LCG (m) of amidships
Fuel	27.0	5.2
Stores and freshwater	19.0	4.7fwd
Baggage	3.5	0.0
Tenders	2.2	17.0 aft

$LCB = 1$ m abaft amidships.

Displacement $= 2643$ tonnes.

When one of the classes was launched at Portsmouth, where the water density is 1025 kg.m^{-3}, it took to the water with a mean draught of 3.98 m and a stern trim of 0.6 m (measured between perpendiculars).

Supposing an exactly similar vessel is to be launched at the Clyde, where the water density is 1006 kg. m^{-3}, estimate the draughts at the perpendiculars after launching.

$[T_{AP} = 4.33$ m, $T_{FP} = 3.77$ m$]$

9 Examples 9

1. The TPM of a wall-sided ship floating in standard salt water at a draught of 2.75 m is 940. The displacement at this draught is 3335 tonnes, and the VCB is 1.13 m below the waterline. The VCG of this vessel is 3.42 m above the keel. The transverse second moment of the waterplane area is 7820 m^4. Calculate the new values of KB, KG and KM_T when a mass of 81.2 tonnes is placed 3.05 m off centreline on a deck 8.54 m above the keel. Estimate the angle of heel due to this addition.
   $[KB = 1.648$ m, $KG = 3.542$ m, $KM_T = 3.994$ m, angle of heel 9.1°$]$
2. A uniform solid rectangular block of square cross section is floating, in freshwater, with one face horizontal. Show that this condition is unstable if the specific gravity of the block lies between:

$$\frac{1}{3 + \sqrt{3}} \quad \frac{1}{3 - \sqrt{3}}.$$

3. A uniform log whose cross section is an equilateral triangle floats, in freshwater, with its vertex down and the top horizontal. Show that this condition is stable if the specific gravity of the log is greater than 9/16.
4. A uniform circular cylinder of height/diameter ratio 0.8 and specific gravity 0.5 floats in freshwater. Show that the following conditions are unstable:

   a. cylinder floats with its axis vertical
   b. cylinder floats with its axis horizontal.

5. A uniform circular cone floats with its vertex down and the top horizontal. The height of the cone is H and the diameter at the top D. Show that when this cone is floating at a draught T, the position of the transverse metacentre is given as:

$$KM_T = \frac{3}{4}T\left[1 + \left(\frac{D}{2H}\right)^2\right]$$

10   Examples 10

1. A catamaran consists of two asymmetric demi-hulls each 4 m long, with a water-line shape as shown in Fig. 6.

   The equation of the curved part of each demi-hull is parabolic given by,

$$y(x) = 0.2 \left[1 - \frac{x^2}{4} \right]$$

   Calculate the transverse second moment of area of this waterline.
   [1.2471 m^4]

2. An ocean platform is supported on 8 columns (vertical members) and 2 pontoons (horizontal members). The columns are cylindrical with a diameter of 6 m, and the pontoons are square sectioned with dimensions 8 m × 8 m. The longitudinal and transverse spacings are shown in Fig. 7.

   Determine the VCG (above the keel) if the GM_T is to be 3 m when floating at a draught of 40 m.
   [14.49m]

3. A one-man canoe and a two-man canoe, of 95 and 190 kg displacement, respectively, are to be designed so that the transverse metacentre is 200 mm above the waterplane. Obtain the waterline length L, the waterline beam B and the draught T for both designs, when floating in freshwater. Compare their B/T ratios. You may use the following information: $\widehat{m}$ is 10,
   VCB is at $0.375T$ below the waterline,

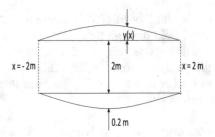

x=-2m   2m   x=2m

y(x)

0.2 m

Fig. 6 Examples 2 5 question 5

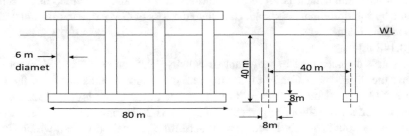

WL

6 m → diamet

40 m

40 m

8m

80 m

8m

Fig. 7 Examples 5 question 6

J_T is $0.045.LB^3$,

C_B is 0.45.

[One-man canoe 4.56 m × 478 mm × 97 mm; two-man canoe 5.75 m × 567 mm × 130 mm]

4. Estimate the transverse metacentre height GM_T for a new class of container ships using the following information: $L = LBP = 270$ m, $B = 35$ m, $T = 11.5$ m, $\Delta = 75000$ tonnes, $C_W = 0.795$, $KG = 10.75$ m, $\rho = 1.025$ tonne.m^{-3}. The waterplane is symmetric about amidships ($x = 0$) and is defined mathematically by:

$$y(x) = \frac{B}{2}\left[1 - \left(\frac{2x}{L}\right)^n\right]$$

The VCB is estimated using Morrish's formula.

[4.14 m]

5. A double bottom tank extending across the ship has the following free surface half-breadths (m), given at equal spacings along the tank:

8.82, 8.00, 6.93, 5.65, 4.24, 2.84 and 1.53.

The tank is 24 m long, and the ship has a displacement volume of 17000 m^3.

Calculate the loss of GM_T due to free surface effect if this tank contains water ballast of same density as the water the ship is floating in.

[0.228 mm]

11 Examples 11

1. $GM_T = 1$ m for a cargo liner of $50MN$ displacement. A crane lifts a weight of $0.2MN$ from a cargo hold through a vertical height of 10 m. The height of the crane's jib is 20 m above this cargo hold, and the radius of the jib is 10 m. Calculate the reduction in GM_T caused by the lifting of the weight. Calculate the approximate angle of heel caused by turning the crane through 30°, assuming it is initially aligned fore and aft.

[[0.08 m], 1.36°]

2. A ship of 10000 tonnes displacement is floating in dock water of density 1024 kg.m^{-3}. This ship is carrying oil of specific gravity 0.84 in a double bottom tank. This rectangular tank is 25 m long and 15 m wide and is divided at its centre line by a longitudinal bulkhead. Calculate the loss of GM_T due to this tank being partially filled.

[0.148 m]

3. A ship's displacement in sea water of density 1025 tonne. m^{-3} is 5100 tonnes, and the distances of Centre of Gravity and transverse metacentre from the keel are $KG = 4$ m and $KM_T = 4.8$ m. A double bottom tank is full of freshwater in the upright condition. This rectangular tank is 20 m long, 6 m wide, 1 m deep, and its bottom is at keel level. The inboard longitudinal bulkhead, which forms one side of this tank, is on the ship's centre line. Assuming KM_T is unchanged,

Table 4 Ship displacement versus draught

Level keel draught (m)	4.25	4.00	3.75	3.5
Displacement (tonnes)	3560	3300	3050	2810
KM_T (m)	5.22	5.40	5.62	5.76

  calculate the heel angle after 60tonnes of freshwater is consumed.
  [2.9°]
4. The underside of the keel of a ship just touches the top of the docking blocks
  when the level keel draught is 4.25 m. The following information in Table 4 is
  given as the water level falls:
  Given that KG is 4.57 m, calculate the transverse metacentre height GM_1 (see
  notes for definition) as the water level falls and, hence, obtain the draught at which
  the equilibrium is neutral.
  [3.524 m]
5. A ship with $LBP = 120$ m, whilst floating at a level keel draught of 7 m, lodges
  on a rock at a point 5 m abaft the forward perpendicular. The ship rests, whilst the
  tide falls 1.2 m before being towed off. The initial displacement is 12400 tonnes,
  and the corresponding GM_T is 0.24 m.
  The following information is available for the range of draughts that need to be
  considered:
  $TPC = 18.4$
  $MCT = 140$ tonne.m/cm
  LCF is at amidships, and KM_T is 7.12 m and assumed unchanged.
  Calculate the draughts at the perpendiculars just before the ship is towed off and
  state whether the ship remained upright or not.
  [$T_{AP} = 7.728$ m, $T_{FP} = 5.717$ m, No]

12   Examples 12

1. A barge of constant rectangular cross section is 60 m long, 8.5 m wide and floats
  in standard salt water at a mean draught of 3.4 m. It is initially unstable and has
  an angle of loll of 6°. A mass of 0.3 tonnes is moved across the deck through
  a distance of 6 m towards the edge nearer to the water. Obtain the additional
  inclination caused.
  [1.9°]
2. The variation of righting lever with heel angle for a cargo vessel, $\Delta = 10000$
  tonnes, is as follows in Table 5:
  Determine the maximum GZ and the Range of Stability.
  [approx. 0.54 m and 83°]

Table 5 Righting levers against heel angle

ϕ (deg)	0	15	30	45	60	75	90
$GZ(m)$	0.0	0.275	0.515	0.495	0.330	0.120	−0.100

Table 6 Ship angle of heel

ϕ (deg)	0	10	20	30	40	50	60	70	80	90
$GZ_{as}(m)$	0.0	0.635	1.300	1.985	2.650	3.065	3.145	3.025	2.725	2.230

Determine the possible angle(s) of heel if a mass of 500 tonnes is moved 10 m across the deck from port to starboard.
[approx. 25° and 73°]
Estimate the Range of Stability if the Centre of Gravity is raised by 0.25m.
[approx. 67°]

3. A ship of $50MN$ displacement has a KG of 6.85 m. The following values in Table 6 were obtained from a set of cross curves of stability, at the above displacement, created using an assumed value for the vertical position of the Centre of Gravity $KG_{as} = 5$ m:
   Calculate the dynamic stability and the potential energy stored against the righting moment to a heel angle of 80°.
   [84.315° m, 73.58M J]

4. A box-shaped vessel of length 65 m, beam 10 m and depth 6 m floats upright at a level keel draught of 4 m in standard salt water and has $GM_T = 0.6$ m.
   Calculate the potential energy stored against the righting moment to a heel angle of 20°.
   [107.2 tonne.m]

13 Examples 13

1. A box-shaped vessel of length 65 m, beam 10 m and depth 6 m floats upright at a level keel draught of 4 m in standard salt water and $KG = 3.0$ m. The vessel has a forepeak compartment 5 m long, 10 m wide and extending to the full depth. Find the new draughts at the perpendiculars if this compartment is bilged, assuming a permeability of 0.9.
   [$T_A = 3.339$ m, $T_F = 5.399$ m]

2. A box-shaped vessel of length 64 m, beam 10 m and depth 6 m floats upright at a level keel draught of 5 m in standard salt water and $KG = 3.0$ m. The vessel has a forward compartment 6 m long and 10 m wide which extends from the keel to a

Table 7 Launching problem

Travel (m)	125	130	135	140	145	150	155	160	165
Buoyancy (MN)	36.27	40.39	44.63	48.96	53.37	57.19	58.30	59.41	60.51
LCB abaft LCG (m)	9.7	8.5	7.4	6.3	5.4	4.4	3.0	1.7	0.4

height of 3.5 m. Find the new draughts at the perpendiculars if this compartment is bilged, assuming a permeability of 0.25.

$[T_A = 4.858$ m, $T_F = 5.306$ m]

3. A box-shaped vessel of length 120 m, beam 18 m and depth 10 m floats upright at a level keel draught of 8 m in standard salt water. The deck has a parabolic sheer forward of 3 m. Estimate, to a first approximation, the centre and length of a compartment which, when damaged and open to sea, would cause this vessel to float at a waterline tangent to the deck at a point 20 m forward of amidships. Assume that the compartment permeability is 0.85. Please note that the parabolic sheer curve (forward) $z_S(x)$ is defined as follows:

$$z_S(x) = s \left(\frac{2x}{L} \right)^2 \qquad\qquad 0 \le x \le 0.5L$$

with reference to amidships deck level. In this expression L and s denote the length and amount of forward, sheer respectively.

[Centre 24 m forward of amidships and length 22.5 m approx.]

4. A cargo ship has a launching weight of $60.82MN$, and the fore poppet is 68.7 m ahead of the LCG. The Centre of Gravity of the vessel passes the way end after 117 m travel. The following information is given in Table 7:

Calculate the least moment against tipping, the distance travelled when the stern lifts, the maximum fore poppet load and comment on whether the fore poppet drops passing the way end.

[approximately 35.7 MNm, 150m, 3.63 MN; No]

Printed in the United States
By Bookmasters